EARTHSHAKING

SCIENCE

EARTHSHAKING

SCIENCE

What We Know (and Don't Know)
about Earthquakes

Susan Elizabeth Hough

PRINCETON UNIVERSITY PRESS

Princeton and Oxford

Copyright © 2002 by Princeton University Press

Published by Princeton University Press, 41 William Street,
Princeton, New Jersey 08540
In the United Kingdom: Princeton University Press, 3 Market Place,
Woodstock, Oxfordshire OX20 1SY

Second printing, and first paperback printing, 2004
Paperback ISBN 0-691-11819-1

The Library of Congress has cataloged the cloth edition of this book as follows

Hough, Susan Elizabeth, 1961–
    Earthshaking science : what we know (and don't know) about earthquakes /
Susan Elizabeth Hough.
        p.   cm.
    Includes bibliographic references and index.
    ISBN 0-691-05010-4 (cloth : alk. paper)
    1. Earthquakes.   I. Title.
    QE534.3 .H68   2002
    551.22—dc21

                                                    2001028751

British Library Cataloging-in-Publication Data is available

This book has been composed in AGaramond and Bluejack by
Princeton Editorial Associates, Inc., Scottsdale, Arizona

Printed on acid-free paper. ∞

pup.princeton.edu

Printed in the United States of America

10 9 8 7 6 5 4 3

FOR

Lee

CONTENTS

PREFACE

In May of 1906, the *San Francisco Chronicle* published a letter from a man named Paul Pinckney, which included the following passage:

> The temblor came lightly with a gentle vibration of the houses as when a cat trots across the floor; but a very few seconds of this and it began to come in sharp jolts and shocks which grew momentarily more violent until the buildings were shaking as toys. Frantic with terror, the people rushed from their houses and in so doing many lost their lives from falling chimneys or walls. With one mighty wrench, which did most of the damage, the shock passed. It had been accompanied by a low, rumbling noise, unlike anything ever heard before, and its duration was about one minute. . . .
>
> It was not until the next day that people began to realize the extent of the calamity that had befallen them. Then it was learned that not a building in the city had escaped injury in greater or less degree. Those of brick and stone suffered most. Many were down, more were roofless, or the walls had fallen out, all chimneys gone, much crockery, plaster, and furniture destroyed.[1]

The earthquake-savvy reader might think that these passages describe the great San Francisco earthquake of April 18, 1906, but they do not. For although the 1906 event inspired the above account, Pinckney was, in fact, describing the earthquake he had experienced twenty years earlier, as a young boy in Charleston, South Carolina.

Today, everyone knows about the San Francisco earthquake, and everyone knows that California is earthquake country. Prior to 1906, however, earthquakes were anything but a California problem. A powerful, prolonged sequence of events—including three with magnitudes at least as large as 7—had rocked the middle of the continent over the winter of 1811—1812. Later that century, the Charleston earthquake had devastated coastal South Carolina and produced perceptible shaking over half of the United States. Its magnitude would also eventually be estimated at higher than 7—nearly as large as the

1999 Izmit, Turkey, earthquake, which claimed more than seventeen thousand lives and left hundreds of thousands homeless.

As a society we recognize the importance of studying (political) history: those who fail to understand the mistakes of the past are condemned to repeat them. Where geologic history is concerned, however, events are out of our control. There is no question that events will repeat themselves. And therein lies the imperative not to control events but rather to understand them, and respond appropriately.

We cannot predict earthquakes; we may never be able to do so. In fact, many Earth scientists now argue strongly that earthquake prediction research should not be a high priority among competing demands for funding, that money is better spent pursuing goals we know to be both attainable and beneficial.

At present, those goals are, broadly, twofold: to quantify the long-term expected *rate* of earthquakes and to predict the *shaking* from future earthquakes. Both goals are inevitably parochial in their focus, for although much can be gained from general investigations of earthquake processes and shaking, earthquake hazard in any region will be shaped critically by the geologic exigencies of that area. In the Pacific Northwest, the process of subduction, whereby an oceanic plate sinks beneath the continental crust, controls hazard. In California, earthquake hazard reflects the lateral motion of two tectonic plates that slide past each other. In the central and eastern United States, the processes that give rise to earthquakes remain somewhat enigmatic, yet history tells us that they are no less real. We know beyond all shadow of doubt that they are real; voices such as that of Paul Pinckney tell us so.

Pinckney's compelling tale included commentary on the state of mind of San Franciscans, who, in 1906, despaired of ever rebuilding their great city, "They forgot the fate of Charleston, so quickly does the mind leap from the events of the yesterdays, however shining, to the more engaging problems which each new day presents."

In 1906, Pinckney had a perspective that other residents of San Francisco lacked, the perspective of having watched a city rebuild itself from ashes. Terrible as earthquakes may be, they shape but do not define society. In the face of natural disasters, humankind has a history of resilience and perseverance. As Pinckney recounted of his boyhood home, "Four years later in 1890, the only visible evidence of this great destruction was seen in the cracks which remained in buildings that were not destroyed. A new and more beautiful, more finished city had sprang up in the ruins of the old."

The question, then, is not whether earthquakes will destroy us in any collective sense but rather what price they will exact from us. How many buildings will be reduced to rubble? How much business interrupted? How many lives lost? Unlike geologic events themselves, the answers to these questions are within our control. By studying earthquake history—both its voices and its geologically preserved handiwork—and earthquake science, we can quantify the hazard and then bring our engineering wherewithal to bear on the task of designing our buildings accordingly.

Tragically, Paul Pinckney's words have largely been forgotten. In 1906 he wrote with authority and without equivocation, "I do not hesitate to assert that the temblor which wrecked Charleston was more severe than that of [the San Francisco earthquake], and in relative destruction, considerably worse."

Yet as society and the Earth science community of the twentieth century strove to understand earthquake hazard and design building codes with adequate earthquake provisions, the seismic history of Charleston, along with that of the rest of the central and eastern United States—and even parts of the West—slipped quietly into obscurity. As the 1906 San Francisco and 1933 Long Beach earthquakes solidified our sense of California as earthquake country, the construction of brick and stone buildings—which Pinckney observed to have suffered disproportionately in 1886—continued apace in Charleston and elsewhere.

Where earthquake safety is concerned, the infrastructure we have is not the infrastructure we want. Vulnerabilities remain, both in California and elsewhere. The imperative is daunting but clear: quantify the hazard, develop the building codes, and strengthen the structures that predate modern earthquake codes. These tasks are, of course, enormously expensive, costing trillions of dollars, and thus are difficult to swallow when the problem is so easy to ignore. As crapshoots go, it is perhaps not a bad bet that no damaging earthquake will occur in any one person's lifetime in any particular place. Collectively, however, the bet cannot be won. Damaging earthquakes *will* strike in the future; that much we can predict. They will moreover strike in unexpected places. When, say, a magnitude 6.5 or 7 temblor strikes New York City or Washington, D.C., or Salt Lake City, what will the richest country on Earth say to its citizenry then?

Paul Pinckney is but one of thousands who made themselves part of the tapestry of earthquake history, the tapestry woven by individuals who did what they could to document and describe the spectacular events they had witnessed. Over the century that followed the publication of Pinckney's account, voices such as his were joined by those of Earth science professionals who endeavored to investigate and understand earthquakes with sophisticated tools and methods. Although our understanding of earthquake science remains far from complete, these voices already unite in a chorus of elegance and power. Their impact—their legacy—depends entirely on the extent to which they are heeded.

The translation of scientific knowledge into public policy requires that the lessons of the former be understood broadly outside of the scientific community, both by the public and by policymakers. *Earthshaking Science* was born of a single "Aha!" moment related to this reality. After reading a succession of newspaper articles in the aftermath of the 1994 Northridge, California, earthquake, I realized that although they were generally good, their technical accuracy was sometimes inconsistent. Then it hit me: these articles are all that the public sees. For Earth science professionals, misleading and inaccurate articles might be frustrating, but for those for whom they are virtually the only source of information, such articles cannot result in anything but confusion.

Within five years of the 1994 quake, the landscape had changed for the better thanks to the emergence of the World Wide Web. Because both journalists and the public were able to turn to the Web for direct scientific information and research results, the level of discourse was significantly elevated. In one memorable phone call in the aftermath of the 1999 Hector Mine earthquake, a reporter from a newspaper in a small desert town told me that she had almost finished her story but just wanted to check one fact. The seismic instruments that were installed in their local area, were they analog or digital? My answer was brief and to the point, and I hung up the phone smiling at the depth of knowledge reflected by the seemingly simple question.

Unfortunately, the anarchy that is one of the Web's greatest strengths is also, at times, its greatest weakness. With abundant earthquake-related information now only a mouse-click away, separating the wheat from the chaff can be a daunting challenge. Especially for the subject of earthquake prediction, one cannot judge an e-book by its cover. That is, a Web site that looks professional and sounds technically sophisticated might nevertheless be espousing arguments considered total drivel by the entire mainstream Earth science community.

Earthquake-related issues—including earthquake prediction—are critically important both to those who live in earthquake country and, sometimes, to those who don't think earthquake hazard is a concern in their neck of the woods. And issues such as prediction and hazard assessment are complex, but not impenetrably so. The first goal of this book is not only to impart specific information but also to develop a framework for understanding, to develop an earthquake science thresher, if you will, with which future chaff can be recognized and dealt with appropriately.

Earthshaking Science might have been born of a perceived need for clearly presented basic information, but its second goal involves something more. Those who devote their lives to scientific research are fortunate to share their profession with colleagues who are passionate and excited about their work. Scientists are, almost by definition, fascinated by their field of interest, be it animal, vegetable, or mineral.

Earth scientists are no exception. It's easy to tell when an Earth science convention is in town; you'll see scores of conference-goers not in standard business attire but rather in blue jeans or khakis, wearing footwear designed for off-road travel. For a fascination with the planet Earth is, inevitably, more than academic. The ranks of Earth scientists are well populated with avid hikers, campers, rock hounds, rock climbers, and so on. Such interests are by no means the exclusive purview of Earth scientists, of course; it is possible to feel passionate about the earth without understanding and appreciating the Earth sciences. Yet appreciation and understanding, which cannot help but strengthen one's passion, are more accessible than they may seem. Most Earth science concepts—even relatively complicated ones—can be explained and understood without technical jargon and mathematical rigor. And understanding, the scientist cannot help but believe, begets appreciation.

The second goal of this book is, therefore, not only to explain scientific concepts but also to communicate the passion and excitement associated with earthquake science. *Earthshaking Science* is not just about what we know but also about how we have come to know it and about why this knowledge is truly extraordinary. The body of knowledge that is modern earthquake science is a remarkable one, gleaned over the ages from inspiration and perspiration alike, from centuries of revolutionary and evolutionary science. Think, for a second, about some of the things we know: the age of the earth, the inner structure of the earth, the movement of the continents over hundreds of millions of years. Like so many things, including the increasingly sophisticated technology that now surrounds us, collective knowledge is easily taken for granted. But to lose sight of the magnitude of past achievement is to miss embracing and appreciating the very best parts of humanity, the very best parts of ourselves. Standing on the shoulders of giants might enable us to see a long way, but recognizing where we stand requires us to stand that much taller and set our sites that much farther ahead.

For those who come to this book looking for basic information with which to make informed decisions about earthquake-related issues, I hope you find these pages clear and useful. But I hope you also find a little more than you were looking for. Earthquake science amazes me; I hope it will amaze you, too.

ACKNOWLEDGMENTS

Because this book focuses on recent developments in earthquake science, most of the scientists whose work is discussed in this book are my colleagues. I am proud to be part of this remarkable community and happy to count many of these talented and extraordinary individuals among my friends. Any attempt to name them all would exceed the space limitations of this short section, and I would inevitably end up mortified at having forgotten someone. I owe my biggest debt of gratitude to the people whose names are sprinkled through the pages of this book: my teachers, my colleagues, my friends.

The list of colleagues in my immediate neighborhood, the Pasadena office of the U.S. Geological Survey and the Caltech Seismology Lab, is shorter and more amenable to enumeration. I therefore acknowledge the following researchers, all of whom conspire to create a work environment that is not only dynamic but also fun: Don Anderson, Rob Clayton, Bob Dollar, Ned Field, Doug Given, Egill Hauksson, Tom Heaton, Don Helmberger, Ken Hudnut, Kate Hutton, Lucy Jones, Hiroo Kanamori, Katherine Kendrick, Nancy King, Aron Meltzner, Sue Perry, Stan Schwarz, Kerry Sieh, Joann Stock, David Wald, Lisa Wald, Brian Wernicke, Bruce Worden, and Alan Yong. Good science doesn't happen in a vacuum, and, seismologically speaking, the corner of California and Wilson Avenues is about as far from a vacuum as one can get without being in a black hole.

I also acknowledge a small number of colleagues who helped make this book happen: Yuehua Zeng, Jim Brune, Katsuyuki Abe, and Kazuko Nagao for graciously providing figures or photographs; Paul Richards for constructive criticism; and Jerry Hough and Chris Scholz for helpful discussions about book-writing in general. I am also grateful to the Southern California Earthquake Center for making available the computer resources that enabled me to put together many of the figures.

This book has been shaped by three editors with whom it has been a pleasure to work: Kristin Gager, Joe Wisnovsky, and Howard Boyer. I am indebted to all of these individuals for editing that helped craft each chapter and the book as a whole. I also thank the staff of Princeton Editorial Associates, whose exemplary copy editing banished the remaining kinks from the manuscript.

In writing this book I was also fortunate to have had assistance from individuals closer to home. Many of the figures in this book—the best ones—are the handiwork of my son Joshua, who was a better graphic artist at twelve years of age than his mother will ever be. Additionally, all three of my kids—Joshua, Sarah, and Paul—served as the primary focus group that helped me come up with the title for this book. And the book is dedicated to my husband, Lee Slice, for reasons that any parent who has ever written a book will understand.

EARTHSHAKING

SCIENCE

ONE THE PLATE TECTONICS
REVOLUTION

Who can avoid wondering at the force which has upheaved these
mountains.

–CHARLES DARWIN, *Voyage of the Beagle*

Our fascination with earthquakes likely dates back to the dawn of human
awareness, but efforts to understand them were doomed to failure prior to the
1960s. Although key aspects of seismology were understood before this time,
seeking a deep understanding of earthquakes in the context of forces that shape
our planet was, for a long time, a little like trying to fathom heart attacks without knowing anything about the body's circulatory system. Prior to the plate
tectonics revolution of the 1960s, humans had struggled for centuries to understand the earth, to reconcile its obviously restless processes, such as seismicity and vulcanism, with its seemingly ageless geology.

This book must therefore begin at the beginning, with an exploration of the
theory that provides the framework within which modern earthquake science
can be understood. We know this theory by a name that did not exist prior to
the late 1960s—the theory of plate tectonics. In a nutshell, the theory describes
the earth's outermost layers: what their components are and, most critically for
earthquake science, how the components interact with one another on our dynamic planet.

THE HISTORY OF A THEORY

Geology is by no means a new field; indeed, it is one of the classic fields of scientific endeavor. By the third century A.D., Chinese scientists had learned how
to magnetize pieces of iron ore by heating them to red hot and then cooling them

in a north—south alignment. Such magnets were widely used as navigational compasses on Chinese ships by the eleventh century A.D. Important discoveries in mineralogy, paleontology, and mining were published in Europe during the Renaissance. Around 1800, an energetic debate focused on the origin of rocks, with a group known as Neptunists arguing for an Earth composed of materials settled from an ancient global ocean, while an opposing school, the Plutonists, posited a volcanic origin for at least some rocks. Investigations of rock formations and fossils continued apace from the Renaissance on. In 1815, geologist William Smith published the first maps depicting the geologic strata of England.

The introduction of a geologic paradigm that encompassed the earth as a whole, however, required nothing short of a revolution. Revolutions don't come along every day, in politics or in science, and the one that changed the face of the Earth sciences in the 1960s was an exciting event. Plate tectonics—even the name carries an aura of elegance and truth. The upshot of the plate tectonics revolution—the ideas it introduced to explain the features of the earth's crust—is well known. The story behind the revolution is perhaps not. It is a fascinating tale, and one that bears telling, especially for what it reveals not only about the revolution itself—a remarkable decade of remarkable advances—but also about its larger context. Revolutionary advances in scientific understanding may be exciting, but they are inevitably made possible by long and relatively dull, but nevertheless critical, periods of incremental, "evolutionary" science.

No schoolchild who looks at a globe can help but be struck by an observation as simple as it is obvious: South America and Africa would fit together if they could be slid magically toward each other. Surely this is not a coincidence; surely these two continents were at one time joined. Upon learning the history of the plate tectonics revolution, one cannot help but wonder why on earth (so to speak) scientists took so long to figure out what any third-grader can see?

One answer is that the idea of drifting, or spreading, continents has been around for a long time. In the late 1500s, a Dutch mapmaker suggested that North and South America had been torn from Europe and Africa. In 1858, French mapmaker Antonio Snider published maps depicting continents adrift. In 1620, philosopher Francis Bacon commented on the striking match between the continents.

Early maps of the world reveal that humankind viewed the earth as a violently restless planet. Such an outlook, which came to be known as *catastrophism* in the mid—nineteenth century, seemed only natural to people who sometimes witnessed—but could not begin to explain—the earthquakes, volcanoes, and storms that provided such compelling and unfathomable displays of power. Such ideas were also consistent with, indeed almost a consequence of, prevailing Western beliefs in a world inexorably shaped by catastrophic biblical events.

By the mid-1800s, however, the paradigm of catastrophism, which described an Earth shaped primarily by infrequent episodes of drastic change, had given way to a new view of the planet, first proposed by James Hutton in 1785 and later popularized by Charles Lyell, known as *uniformitarianism*. Based on recognition of the nearly imperceptible pace of geological processes and the earth's considerable age, the principle of uniformitarianism holds that geologic processes have always been as they are now: at times catastrophic, but more often very gradual. If continents are not being torn apart now, proponents of the new school of thought argued, how could they have been ripped from stem to stern in the past?

By the nineteenth century, moreover, the technical sophistication of the scientific community had grown, and scientists became more concerned than they had been in the past with understanding physical *processes*. The mid—nineteenth century was an extraordinary time for science. Charles Darwin's *On the Origin of Species* was published in 1859 (Sidebar 1.1). Seven years later, Gregor Mendel laid out his laws of heredity, which established the basic principles of dominant and recessive traits. At nearly the same time, Louis Pasteur's germ-based theories of disease gained wide acceptance. Advances were by no means restricted to the biological sciences. In 1873, James Maxwell published the four equations that to this day form the backbone of classic electromagnetism theory.

In this climate of growing scientific sophistication, advocating a theory of continental drift without providing a rigorous physical mechanism for the phenomenon became untenable. If the continents moved, how did they move? In 1912, a German meteorologist named Alfred Wegener presented the basic tenets of continental drift in two articles. He introduced a name for the supercontinent that existed prior to the break-up that separated Africa from South America, a name that remains in use today: Pangaea. Wegener's ideas were scarcely flights of unscientific fancy; they were instead based on several types of data, primarily from paleobotanical and paleoclimatic investigations. Wegener pointed to evidence that tropical plants once grew in Greenland and that glaciers once covered areas that are at midlatitudes today; he proposed continental drift as a mechanism to account for these observations.

When challenged over the ensuing decades to produce a physical mechanism to go with the conceptual one, Wegener put forward the notion that the continents plow their way through the crust beneath the oceans. The image conveyed, that of continental barges adrift on the oceans, had intuitive appeal. The mechanism, however, was easily dismissed by geophysicists, who understood enough about the nature of the earth's crust to know that continents could not push their way through the floor of the oceans without breaking apart.

For nearly two decades the debate raged on. Tireless and determined by nature, Wegener published a book and many papers on his theory. Yet he was not to see vindication during his own lifetime. Tragically, Wegener died from exposure during a meteorologic expedition to Greenland in 1930. In the annals of Earth science, few deaths have been quite as badly timed as Alfred Wegener's at the age of fifty. He died just as oceanic research vessels were beginning to acquire high-quality seafloor topography (bathymetric) data that would provide scientists a vastly better view of the character of the ocean floor and pave the way for the revolution to begin in earnest.

WEGENER'S HEIRS

Harry Hammond Hess, born in 1906, was a scientist whose timing and luck were as good as Wegener's were bad. A geologist at Princeton University in the late 1930s, Hess was well poised to build upon Wegener's continental drift hypothesis, as well as a hypothesis, proposed in 1928 by British geologist Arthur Holmes, that material below the crust circulated, or convected, much like water in a boiling pot (but far more slowly). Hess, a member of the U.S. Naval Reserves, was pressed into service as captain of an assault transport ship during World War II. Military service might have been nothing more than an unfortunate interruption in a sterling research career, but Hess was not a man to let opportunity pass him by. With the cooperation of his crew, he conducted echo-sounding surveys to map out seafloor depth as his warship cruised the Pacific. Marine geophysics is an expensive science, primarily because of the high cost of operating oceanic research vessels. There is no telling how long it might have taken the geophysical community to amass the volume of seafloor topography data that Hess collected while he happened to be in the neighborhood, but the data set was a bounty on which Hess himself was able to capitalize almost immediately.

Back at Princeton after the war, Hess turned his attention to the character of the ocean floor as revealed by his extensive surveys. Struck by the nature of the long, nearly linear ridges along the seafloor—the so-called mid-ocean ridges—away from which ocean depth increased symmetrically, Hess prepared a manuscript presenting a hypothesis that would become known as seafloor spreading. A first draft prepared in 1959 circulated widely among the geophysical community but was not published. Hess's ideas were met with skepticism and resistance, just as Wegener's theories had been earlier. To argue that the structure of the ridges and ocean basins implied a mechanism of spreading was little different from arguing that the striking fit between Africa and South America implied continental drift. Both hypotheses failed to provide a physical mechanism, and both were inherently descriptive.

Hess, however, had his remarkable data set and the framework for a geophysical model. In 1962, he published a landmark paper titled "History of Ocean Basins," which presented a mechanism for seafloor spreading. Hess's model described long, thin blades of magma that rise to the surface along the mid-ocean ridges, where they cool and begin to subside as they get pushed away bilaterally as more oceanic crust is created. Millions of years after being created at a mid-ocean ridge, the crust encounters trenches along the ocean rim, where the crust then sinks, or *subducts,* descending back into the earth's mantle in the granddaddy of all recycling schemes. These trenches, also imaged by Hess's seafloor surveys, were a critical element of the hypothesis because they explained how new crust could be created continuously without expanding the earth. Some of Hess's ideas about subduction were incomplete, however, and were superseded by later studies and models.

Still, Hess's 1962 paper was remarkable for its prescience and its insights into physical processes. Although it was what scientists consider an "idea paper" (that is, one whose hypotheses are consistent with but cannot be proven by the data in hand), the 1962 paper was distinguished from its predecessors by its presentation of nineteen predictions derived from the model. Ideas may be the seeds of good science, but testable hypotheses are the stuff of which it is made. Scientific hypotheses are accepted not when they offer a satisfactory explanation of empirical observations but when they allow predictions that are borne out by new, independent data. Of Hess's nineteen predictions, sixteen would ultimately be proven correct.

The true Rosetta stone of plate tectonics—the key that won over a skeptical community—involved something considerably simpler than hieroglyphics: the geometry of the linear, magnetic stripes on the ocean floor. That the oceanic crust was magnetic came as no surprise; by the 1950s, scientists knew that it was composed predominantly of basalt, an iron-rich volcanic rock well known for its magnetic properties. Basalt was moreover known to serve as a permanent magnetometer, locking in a magnetism reflecting the direction of the earth's magnetic field at the time the rock cooled (not unlike the ancient Chinese magnets).

As early as 1906, scientists investigating the magnetic properties of the earth's crust recognized a tendency for rocks to fall into one of two magnetic orientations: either aligned with the earth's present-day magnetic field or aligned in the diametrically opposite direction (normal polarity and reversed polarity, respectively). The earth's magnetic field is similar the field generated by a bar magnet with its north end nearly aligned with the geographic North Pole. Yet the earth's field is the result of a more complex, dynamic process: the rotation of the planet's fluid, iron-rich core. Although the process gives rise to a field that appears fixed on short timescales, scientists have known for centuries that the earth's magnetic field is dynamic and evolving. At a rate fast enough to be measured even by the slow pace of human investigations, the magnetic field drifts slowly westward at a rate of approximately 0.2 degrees per year. Over tens of thousands of years, the field undergoes far more dramatic changes known as *magnetic reversals*. During a reversal, south becomes north and vice versa, apparently in the blink of an eye, at least from a geologic perspective—perhaps over a period of a few thousand years. Basaltic rocks lock in the field that existed at the time the rocks were formed, and they remain magnetized in that direction, insensitive to any future changes in the earth's field.

The subfield of geology known as *geomagnetism*—the study of the earth's magnetic field—has long been an intellectually lively one. The study of earthquakes was hampered for years by the slow pace of seismic events (in Europe especially) and the lack of instrumentation to record them. The earth's magnetic field, by contrast, is considerably more amenable to investigation with simple instruments. Documentation of the magnetic field—its direction and intensity—dates back nearly 400 years, to William Gilbert, physician to Queen Elizabeth I.

Beginning in the 1950s, geomagnetism found its way to the forefront of Earth sciences in dramatic fashion. The journey began, as so many scientific journeys do, serendipitously. When the ocean floor was surveyed with magnetometers designed to detect submarines during World War II, a strange pattern gradually came into focus. The oceanic crust was no mottled patchwork of normal and reversed polarity; it was striped. In virtually every area surveyed, alternating bands of normal and reversed polarity were found (Figure 1.1).

A report by U.S. Navy scientists in 1962 summarized the available magnetic surveys of oceanic crust. Just a year later, British geologists Frederick Vine and Drummond Matthews proposed a model to account for the observations in the navy report. They suggested that the oceanic crust records periods of normal and reversed magnetic alignment, in the manner that had been documented earlier for continental crust. To interpret the spatial pattern, Vine and Matthews applied an equation from high school physics: velocity equals distance divided by time. In September of 1963, the team of scientists published a paper in the journal *Nature* in which they proposed that the magnetic stripes resulted from the generation of magma at mid-ocean ridges during alternating periods of normal and reversed magnetism; the scientists' proposal was consistent with the predictions from Hess's seafloor-spreading hypothesis.

Vine's and Matthews's work paralleled that of another scientist working independently, Lawrence Morley of the Canadian Geological Survey. Such coincidences are neither unusual nor remarkable in science. The collective wisdom of a scientific field is built slowly, not so much in leaps of extraordinary genius as in a series of steady steps taken by individuals who possess both a thorough understanding of the current state of knowledge and the talent to take the next logical step. In the case of ocean floor magnetism, the 1963 publication of Vine and Matthews was actually preceded slightly by a paper that Morley had submitted twice, a paper that had been rejected twice, once by *Nature* and once by one of the seminal specialized Earth sciences journals, the

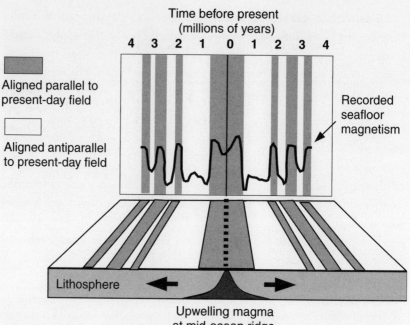

Time before present
(millions of years)

4 3 2 1 0 1 2 3 4

Aligned parallel to
present-day field

Aligned antiparallel
to present-day field

Recorded
seafloor
magnetism

Lithosphere

Upwelling magma
at mid-ocean ridge

Figure 1.1. Magnetic striping of oceanic crust compared to the magnetic record established from continental crust. Dark and light bands correspond to magnetic alignments that are parallel and antiparallel to the present-day field direction, respectively. The recorded seafloor magnetism *(dark line in top panel)* reveals the alternating orientations.

Journal of Geophysical Research. In a remark that has become famous in the annals of the peer-review publication process, an anonymous reviewer commented that Morley's ideas were the stuff of "cocktail party conversation" rather than science. Historians of science—and scientists—do generally give Morley his due, however, when they refer to the "Vine-Matthews-Morley hypothesis."

In 1963, Canadian geophysicist J. Tuzo Wilson introduced another "cocktail party" concept that was critical to the development of plate tectonics theory (Sidebar 1.2). Focusing his attention on volcanic island chains like Hawaii, Wilson suggested that the characteristic arc shape of these islands resulted from the passage of the crust over an upwelling of magma that remained stationary in the mantle. Wilson dubbed these upwellings "hotspots," a term (and concept) that has long since been accepted into the scientific lexicon but was so radical in its day that Wilson's paper, like Morley's, was rejected by several major geophysical journals before seeing the light of day in a relatively obscure Canadian geological journal.

Wilson continued to work on critical aspects of plate tectonics theory after 1963, as did Vine and Matthews, who continued their investigations of magnetic stripes. To better understand the time frame for the formation of the stripes, Vine and Matthews looked to the very recent work of U.S. Geological Survey and Stanford scientists Allan Cox, Richard Doell, and Brent Dalrymple. This team had succeeded in using the slow radioactive decay of elements within basalt to construct a history of the earth's magnetic field dating back 4 million years. By dating rocks from around the world and measuring their magnetization, the scientists produced a 4-million-year timeline indicating periods (epochs) of normal and reversed magnetic field alignment. Comparing the record from continental crust with established observations of magnetic striping in the oceans, Vine and Matthews showed that the two were remarkably consistent if one assumed a seafloor-spreading velocity of a few centimeters a year. Their work was published, again in *Nature,* in 1966.

By the mid-1960s, the revolution was in full swing; the geophysical community was ignited by the exciting ideas that circulated like a firestorm in many, if not quite all, Earth science departments. Some of the ideas might have been old, but the data and theories in support of them were shiny and new. Like so many puzzle pieces, other aspects of plate tectonics were pieced together in a series of seminal papers published in the latter half of the decade.

In a 1965 paper—this one accepted by *Nature*—Tuzo Wilson presented an explanation for the great oceanic transform faults. Long, linear fractures running perpendicular to ocean ridges, these transform faults represented a paradox until Wilson, a scientist fond of using simple cardboard constructions to explore complex geometrical ideas, showed how they fit neatly into a seafloor-spreading model (Figure 1.2).

Although the plate tectonics revolution began at sea, it spread almost instantly to incorporate continental tectonics as well. In 1965, Cambridge University professor Teddy Bullard showed that for continental drift to be a viable theory, the continental plates themselves must be rigid, characterized by very little internal deformation. In 1967, Dan McKenzie and Bob Parker published a paper that worked out the details of rigid plate motion on a spherical earth. A 1968 paper by Jason Morgan of Princeton used the geometry of the oceanic transform faults to show that, like the continents, oceanic plates behave as rigid plates.

At this point in the story it is appropriate to pause and reflect on another great technical revolution of the twentieth century: the computer revolution. With a computer in every classroom and e-mail giving regular ("snail") mail and telephones a run for their money as the preferred means of communication, it is perhaps easy to forget just how new a phenomenon these computing machines are. The first all-electronic computer, ENIAC, made its debut in 1946. It contained eighteen thousand vacuum tubes, could perform a few hundred multiplications per minute, and could be reprogrammed only by manual rearrangement of the wiring. Transistors, which allowed for the development of much smaller computers with no unwieldy vacuum tubes, debuted only in the late 1950s. Silicon chips, which paved the way for the microprocessors we have come to know and love, arrived on the scene two decades later.

Because computer technology was in its infancy in the mid-1960s, making a detailed scientific map was no easy matter, and an early critic of plate tectonics, noted geophysicist Harold Jeffreys, suggested that perhaps a measure of artistic license had been used in the diagrams and mechanical models illustrating plate motions and reconstructions. Geophysicist Bob Parker will cheer-

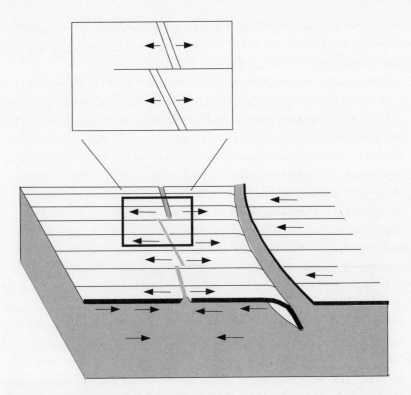

Figure 1.2. Model of transform faults in oceanic crust proposed by J. Tuzo Wilson. Transform faults are depicted by horizontal lines that connect segments of spreading centers, where new crust is created.

fully admit that his contribution to the plate tectonics revolution had less to do with geophysical insight than with his talent for exploiting the newfangled tools of the geophysical trade. Having written a computer program to plot coastlines (Super Map, affectionately named for the then prime minister of England, Harold "Super Mac" Macmillan), Parker teamed up with Dan McKenzie to produce computer-generated three-dimensional reconstructions that bore no trace of artistic license. Even with Super Map, the work was no easy task. McKenzie worked on the reconstructions at Princeton but could not generate maps because the university had no plotter at the time. He tried to make his maps at the Lamont-Doherty Geological Observatory of Columbia University (now the Lamont-Doherty Earth Observatory); the observatory had a plotter but, unfortunately, insufficient computer memory to handle the large (for their day) files. To finally generate the maps for the paper that was published in *Nature* in 1967, McKenzie had to send his files across the ocean to Cambridge University.

Another small handful of landmark papers, perhaps half a dozen, were published between 1965 and 1968 as the pieces of a maturing, integrated theory fell into place. A special session on seafloor spreading was convened at the 1967 annual meeting of the American Geophysical Union; 30 years later the special session would be recognized as a watershed event in the development and acceptance of global plate tectonics theory. Resistance to the theory of plate tectonics persisted through the mid-1960s, with eminent scientists among the ranks of both the latest and the earliest converts. By the late 1960s, however, there was too much evidence from too many different directions; notwithstanding a trickle of continued objection from the most determined quarters, the theory passed into the realm of conventional wisdom (Figure 1.3).

If the pieces of the plate tectonics puzzle had largely fallen into place by 1967 or 1968, the name for this grand new paradigm emerged, curiously enough, rather gradually. Although the papers by McKenzie and Parker (1967) and Morgan (1968) were credited in a 1969 "News and Views" summary in *Nature* as having established the theory of "plate tectonics," neither article had proposed that particular name. McKenzie and Parker perhaps get the close-but-no-cigar award with the title "The North Pacific: An Example of Tectonics on a Sphere" and an opening sentence that read, economically, "Individual aseismic areas move as rigid plates on the surface of a sphere."[4] However, throughout the text they referred to their theory by a name that was not destined to catch on: "paving stone tectonics." Perhaps the earliest official use of the name "plate tectonics" came in the title of a small geophysical meeting convened in Asilomar, California, in 1969 so that participants could present and discuss the exciting new ideas in seafloor spreading and continental drift.

SEISMOLOGY AND PLATE TECTONICS

Seismology might not have been the first subfield of Earth science to arrive at the plate tectonics ball, but it did arrive in time to make its share of critical contributions. In 1968, Xavier Pichon, then a graduate student at Lamont-Doherty Geological Observatory, published a paper that used catalogs of large earthquakes worldwide to help constrain the geometry of the plates. Just a year earlier, another Lamont scientist, Lynn Sykes, had also made a critical seismo-

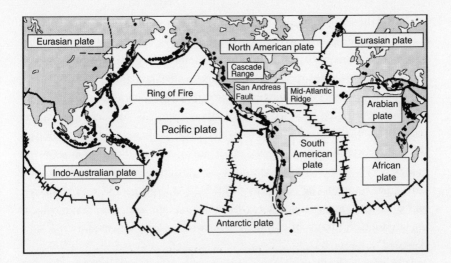

Figure 1.3. The earth's major tectonic plates. The so-called Ring of Fire includes both transform faults such as the San Andreas Fault in California and, more generally, subduction zones around the Pacific Rim. At mid-ocean ridges such as the Mid-Atlantic Ridge, new crust is created. The dots in the figure indicate active volcanoes.

logical contribution to the development of plate tectonics theory. Whereas other seismologists had successfully investigated the forces that were inferred to drive earthquakes, Sykes focused on the inferred fault motions and showed them to be consistent with the motions predicted for Wilson's transform faults.

To understand why earthquake motions do not necessarily reflect driving forces directly, imagine a book on a table. Push on one edge and it will slide across the table in a direction parallel to the direction of the applied force. Now push the book downward into the table and forward. The book will still slide forward, not because that is the direction parallel to the force but because the book is not free to move vertically. The earth's crust behaves in a similar manner when subjected to a driving force. Faults represent pre-existing zones of weakness in the earth's crust, zones along which movement will tend to be accommodated, even if the faults are not perfectly aligned with the driving forces. (The direction of earthquake motions can generally be determined with fewer assumptions than can the direction of the actual driving forces.)

Just as an explosion in the quantity and quality of geomagnetic and topographical data presaged the giant leap in our understanding of oceanic crust, so too did the seismological contributions to plate tectonics theory depend critically on enormous improvements in observational science. Oddly enough, the enmity of nations once again proved to be among the greatest, albeit entirely unwitting, benefactors to science. The Worldwide Standardized Seismograph Network (WWSSN) was launched in the early 1960s to monitor underground nuclear weapons tests and to provide support eventually for a comprehensive ban on all nuclear testing. For the first time, humankind had developed weapons powerful enough to rival moderate to large earthquakes in terms of energy released: both nuclear weapons and earthquakes generate seismic signals large enough to be recorded worldwide. At large distances, earthquakes and large explosions generate what seismologists term *teleseismic waves,* which are vibrations far too subtle to be felt but which can be detected by specially designed seismometers.

It is often difficult to find financial support for scientific inquiry that requires expensive and sophisticated scientific instrumentation. Like commercial electronics, scientific instruments became enormously more sophisticated in the latter half of the twentieth century. Unlike consumer electronics, however, scientific instruments are not subject to mass-market pressures that drive prices down. Prior to the 1960s, a standardized, well-maintained, global network of seismometers would have been a pipe dream for a science that generally runs on a shoestring. Financed not by the National Science Foundation but by the Department of Defense, the WWSSN sparked the beginning of a heyday for global seismology. In the decades following the launch of the WWSSN, data from the network allowed seismologists to illuminate not only the nature of crustal tectonics but also the deep structure of our planet.

The tradition of symbiosis between military operations and academic geophysics continues today. Seismology's contribution to the critical issue of nuclear-test-ban-treaty verification has resulted in substantial support for seismological research and monitoring that would otherwise not have been available.

The military provided yet another boon for geophysics—especially for the study of global tectonics—with the development and implementation of the Global Positioning System (GPS). Initiated in 1973 by the Department of Defense as a way to simplify military navigation, the GPS now relies on dozens of satellites and associated ground-based support. Instrumented with precise atomic clocks, each GPS satellite continuously broadcasts a code containing the current time. By recording the signal from several satellites and processing the pattern of delays with a specially designed receiver on the ground, one can precisely determine the coordinates of any point on Earth. Conceptually, the procedure is nearly the mirror image of earthquake location methods. To locate earthquakes, seismologists use waves generated by a single earthquake source recorded at multiple stations to determine the location of the source.

Within a decade of the inauguration of the GPS, geophysicists had begun to capitalize on its extraordinary potential for geodetic surveys. The dictionary defines *geodesy* as the subfield of geophysics that focuses on the overall size and shape of the planet. Geodesists have historically employed precise ground-based surveying techniques such as triangulation and trilateration. By measuring the transit time of a light pulse or a laser beam between precisely oriented markers on the earth's surface and applying simple geometrical principles, geodesists can determine relative position.

Geodesy has enormous practical applications. We need precise knowledge of location to build roads, to make real estate transactions, and to draw maps, to name but a few applications. Geodetic surveys have contributed to our understanding of the earth's landforms (for example, the size and shape of mountains), to the determination of permanent deformation caused by large earthquakes and volcanic eruptions, and to our knowledge of long-term, slow crustal movements. Efforts to understand slow crustal deformation were hampered for decades by three factors: the intrinsically slow nature of such movements, the imprecision of surveying techniques, and the enormous number of hours required for large-scale geodetic surveying campaigns. The last factor was especially critical because a triangulation or trilateration survey of an area is done by leapfrogging instruments between locations for which there is direct line-of-sight. Covering an area of geophysical interest typically required weeks or months of effort on the part of teams that made their way at a snail's pace, sleeping in mobile encampments.

Although the modern GPS offers significant improvements in both precision and ease of measurement, early geophysical GPS campaigns were arduous. They required teams to observe satellite signals during the limited hours of the day or night when satellite coverage was sufficient to obtain the coordinates of any given location (Sidebar 1.3). By the 1980s, though, satellite coverage had improved and instruments could monitor GPS signals continuously. By the mid-1990s, the geophysical community had begun to implement continuous GPS networks that ran alongside their seismic counterparts. Whereas seismic instruments record ground motions that occur over seconds, minutes,

Geophysicists who participated in the early GPS experiments often recall their experiences with a mixture of humor and affection.
Geodesist Nancy King described measurements taken "with flashlight illumination by people who were bleary-eyed and bone-
cold." She added that "hanging out in vehicles in remote areas at
night also tends to look suspicious," especially at a time when few
people outside the scientific community had heard of GPS. As
King observed, "The police sometimes had a hard time accepting
our explanation that we were out tracking satellites in the wee
hours of the morning because we wanted to study earthquakes."[5]

and sometimes hours, GPS networks record ground motions that occur over
days, months, and years.

GPS data have been used to address a wide range of unresolved questions in
geophysics, including several questions related to earthquake prediction. The
primary geophysical application of these data, however, has been to document
the plates' motion, including their internal deformation. Several decades after
the revolution swept in the grand new ideas, the basic paradigm is clear, but
the devil remains in the details. GPS, a boon to the geophysical community
provided by the defense industry, is one of the most valuable tools for addressing those devilish details.

The symbiotic relationship between the Earth sciences and the military continues to this day. As the twentieth century drew to a close, Earth scientists began to make use of yet another technology originally developed with military
applications in mind: Synthetic Aperture Radar, or SAR. SAR is a technique
that uses reflected microwave energy from satellites to create high-resolution
images of the earth's surface.

In the early 1990s, scientists realized that the differences between two SAR
images of the same region could be used to study earthquakes. This technique,
known as InSAR, was first applied by geophysicist Didier Massonnet and his
colleagues, who studied the 1992 magnitude 7.3 Landers, California, earthquake. In an InSAR image such as the one on the cover of this book, fault displacement can be inferred from a pattern of fringes. Each fringe—a suite of
colors from violet to red, or shades of gray—corresponds to a certain change

in the distance from each point on the ground to the satellite. Typically the change is on the order of a few centimeters per fringe.

The details of InSAR processing and analysis are somewhat complicated, but estimating fault rupture from an InSAR image is basically akin to estimating the age of a tree by counting its rings. A ruptured fault appears as a bull's-eye—or an elongated bull's-eye—in an InSAR image, and the amount of fault motion is given by the number of fringes that make up the bull's-eye.

Unlike GPS measurements, which are available as point measurements from individual stations, InSAR has the potential to image the entire surface of the earth. In some cases, then, it provides far more detailed information than can be obtained using GPS. But GPS is by no means obsolete, as it can provide results in certain areas, such as those with heavy forestation, where InSAR often doesn't work. InSAR also captures only snapshots taken by repeat passes of a satellite, whereas GPS data can be collected continuously. The two techniques are therefore highly complementary, with InSAR providing another nifty (and sometimes delightfully colorful) tool in the Earth scientist's bag of tricks.

THE FRAMEWORK OF PLATE TECTONICS THEORY

By the time the revolutionary dust had settled at the end of the 1960s, a basic understanding of the earth's crust had been established. The "crust" is broken into about ten major plates, each of which behaves for the most part as a rigid body that slides over the partially molten "mantle," in which deformation occurs plastically. The quotation marks around "crust" and "mantle" reflect a complication not generally addressed by the most elementary texts. Geophysicists determine what is crust and what is mantle on the basis of the velocity of earthquake waves. At a depth of approximately 40 kilometers under the continents, wave velocities rise abruptly at a boundary known as the *Mohorovicic discontinuity,* or simply the Moho, which is thought to reflect a fundamental chemical boundary. The Moho separates the crust from the mantle below. Except for those that occur in subduction zones, earthquakes are generally restricted to the upper one-half to two-thirds of the crust, the brittle upper crust. The thickness of the crust ranges from a few kilometers under the oceans to several tens of kilometers for the thickest continents.

Although earthquakes are generally restricted to the crust, the depth of the crust does not coincide perfectly with the depth of the tectonic plates. Continental plates in particular are much deeper, perhaps 70 kilometers on average. Geophysicists know the earth's relatively strong upper layer as the *lithosphere*, only the uppermost layer of which is, in a strict sense, the crust.

Underneath the lithosphere is a layer scientists know as the *aesthenosphere*, which is a zone of relative weakness. Earth scientists categorize layers as weak or strong, often relying on the speed of earthquake waves through a zone as a proxy for its strength. The lithosphere is strong as a unit; wave velocities are high (shear wave velocities of 4.5–5 kilometers/second) over most of its 70-kilometer extent. Below the lithosphere, shear wave velocities drop by about 10 percent, and seismic waves are strongly damped out, or attenuated. Because laboratory results show that zones of partial melting are characterized by slow wave velocities and high attenuation, the aesthenosphere is thought to be a zone of partial melting. That is, the aesthenosphere is not a liquid per se but rather a saturated matrix. The basaltic magma that rises to the earth's surface at the mid-ocean ridges is thought to be derived primarily from the 1–10 percent of the aesthenosphere that exists as a melt.

The weak aesthenosphere, extending to a depth of about 370 kilometers, accounts for the mobility of the solid overriding lithospheric plates. The mantle, as strictly defined, incorporates the aesthenosphere and the solid lower mantle. Deeper still, a chemical and physical boundary marks the transition from the magnesium—iron silicate mantle to the mostly iron core.

The study of the earth's deep interior is a fascinating one and one that is critical to a full understanding of the processes that affect the earth's surface. Many questions remain unanswered. Does the mantle convect, or turn over, as a whole, like one big, slowly boiling cauldron, or does it convect in layers? Where does the magma source for hotspots originate, and how do these features remain fixed in a convecting mantle? What becomes of oceanic crust once it subducts? Do dynamic processes within the mantle help buoy mountain ranges to their present gravity-defying heights?

Sidestepping such questions now to return to the phenomenology of the crust as we understand it—and as it concerns earthquakes—let's focus on the boundaries between the plates. Three types of plate boundaries are defined according to their relative motion: zones of spreading, zones of relative lateral motion (transform faults), and zones of convergence, where plates collide. Simplified examples of faults associated with the three plate-boundary types are shown in Figure 1.4. As already noted, plates pull apart at the mid-ocean ridges, where basaltic magma from the aesthenosphere rises and creates new

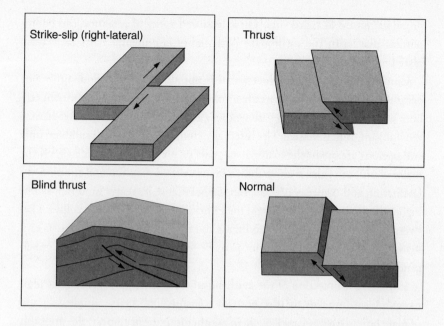

Figure 1.4. The basic types of faulting: *top left*, strike-slip; *top right*, thrust; *bottom right*, normal; and *bottom left*, blind thrust. The different types of plate boundaries are characterized by different types of faulting; strike-slip faulting dominates at transform boundaries, thrust faulting dominates at subduction zones, and normal faulting dominates at zones of spreading.

oceanic crust. Plates converge along subduction zones, where oceanic crust subducts beneath the continents. And plates sometimes slide past one another without any creation or consumption of crust, as with the San Andreas Fault in California and the North Anatolian Fault in Turkey, which produced a devastating magnitude 7.4 (M7.4) event in August of 1999 and a subsequent M7.2 temblor three months later.

Readers will likely not be surprised to learn that the earth is more complicated than the images often depicted in simple drawings. Although mid-ocean ridges are the most conspicuous and active zones of spreading, they are not the only ones. In some cases, deep earth processes conspire to tear a continent apart. The East African rift zone, stretching from Ethiopia south toward Lake Victoria and beyond into Tanzania, is one such zone that is active today. Sometimes continental rifting gets started but then fizzles out, creating what geophysicists regard as failed rifts. These fossilized zones of weakness can be important players in the earthquake potential of continental crust away from plate boundaries.

Convergence is another process that is not always simple. Sometimes one oceanic plate subducts underneath another; and sometimes one continent collides with another, in which case neither can sink because continental crust is too buoyant. If material can't be internally compressed and can't subduct, only two options are open: the material can rise, or it can push out sideways if the material on both sides is amenable to being pushed. Continental collision is a fascinating and complex process, especially because it creates mountains. The highest mountain range on Earth, the Himalayas, is the result of a collision between a once separate Indian land mass and the Eurasian plate. Scientists estimate this collision to have begun perhaps 40 million years ago, and convergence continues to this day.

If the slow convection of the aesthenosphere is the engine that drives plate tectonics, earthquakes are in a sense the transmission. Again, simple diagrams of plate boundaries are inadequate in another important respect: the smoothly drawn boundaries between plates do not accurately represent the complicated geometries and structure of real plate boundaries. An oceanic plate that subducts under a continent is lathered with sediments and possibly dotted with underwater mountains known as seamounts. To descend into an oceanic trench, a plate must overcome significant frictional resistance from the overriding continental crust. Oceanic crust sinks beneath the continents because the former is more dense, but it does not go quietly into the night. Although magma and crustal generation at mid-ocean ridges is relatively continuous, a plate usually stalls out at the other end of the conveyor belt until enough stress accumulates to overcome friction. Oceanic real estate disappears piecemeal, in parcels up to 1,000 kilometers long and 300–500 kilometers deep. By virtue of their enormous area, subduction zones produce the largest earthquakes anywhere on the planet.

Zones of continental convergence are also, not surprisingly, characterized by significant seismic activity. The seismicity associated with the India—Eurasia collision is more diffuse and complicated than that associated with classic subduction zones, but it is no less deadly. Very large earthquakes—the equal of those on the San Andreas Fault and then some—occur over vast regions within Eurasia, including India, Tibet, Mongolia, and mainland China, all a consequence of a collision that began 40 million years ago. The M7.6 earthquake on January 26, 2001, in Bhuj, India, was a consequence of the India—Eurasia collision, and it happened several hundred kilometers away from of the active plate boundary.

Earthquakes are by no means restricted to convergence zones. Crustal generation at mid-ocean ridges is also accompanied by earthquakes. Although generally of modest size, a steady spattering of earthquakes clearly delineates these plate boundaries. At spreading centers, the plates' total motion is greater than the component contributed by the earthquakes: because the spreading process does not involve as much frictional resistance as subduction does, some of the motion occurs gradually, without earthquakes. Along transform faults, however, long-term motion is also accounted for predominantly by the abrupt fits and starts of earthquakes.

The circum-Pacific plate boundaries—sometimes known as the Ring of Fire because of the relentless and dramatic volcanic activity—alone account for about 75 percent of the seismic energy released worldwide. A trans-Asiatic belt, stretching from Indonesia west to the Mediterranean, accounts for another 23 percent or so. That leaves a mere 2 percent of the global seismic energy budget for the rest of the world, including most of the vast interiors of North America, Australia, South America, and Africa. Although 2 percent might not sound like much, it is nothing to sneeze at. The devastating earthquakes that struck midcontinent in the United States near Memphis, Tennessee, in 1811 and 1812 occurred far away from active plate boundaries, in what geologists term *intraplate crust*. Another noteworthy North America event, perhaps as large as magnitude 7, struck Charleston, South Carolina, in 1886. Other events with magnitudes between 6 and 7 have been documented during historic times in the northeastern United States and southeastern Canada.

These intraplate events, which have been attributed to various forces, including the broad compression across a continent due to distant mid-ocean spreading and the slow rebound of the crust upward following the retreat of large glaciers, are considerably less frequent than their plate-boundary counterparts. These events are also potentially more deadly because they strike areas that are considerably less well prepared for devastating earthquakes.

A RETROSPECTIVE ON THE REVOLUTION

Before we explore earthquakes in more detail, it is worth pausing briefly to look back on the plate tectonics revolution as a phenomenon unto itself. Scientific revolutions are indeed uncommon and therefore interesting events; to understand them is to understand the nature of scientific inquiry in general. Although the basic ideas of continental drift might have been obvious, the data and theoretical framework required for the formation of a mature and integrated theory required a technical infrastructure that was unavailable in earlier times.

What do we make of Alfred Wegener? Should we decry the treatment his visionary ideas received and honor him posthumously as the father of plate tectonics? He did, after all, argue tirelessly and passionately for continental drift some 30 or 40 years before it passed into the realm of conventional wisdom. Or should he be dismissed as a nut, having championed ideas he fervently believed even when doing so meant relying on less-than-definitive data and invoking explanations that the experts of the day dismissed as total bunk?

Both arguments have been made in the decades following the 1960s. For those predisposed to viewing the scientific establishment as a monolithic, territorial, and exclusionary entity, Alfred Wegener is nothing short of a poster child. Just look at the derision that his ideas met, the argument goes, because he was an outsider to the field of solid earth geophysics, an outsider who challenged the conventional wisdom of his day.

In the final analysis, however, one must always remember that science is about ideas alone. Having drawn on data from many different sources in support of his ideas, Wegener's contributions to plate tectonics exceed those of Francis Bacon and Antonio Snider. Indeed, by the 1980s, Wegener was frequently credited as being the father of continental drift theory, and rightly so.

But was the treatment of Alfred Wegener in his own time unconscionable? Was the establishment monolithic? Exclusionary? It is easy to conclude that it was, and this conclusion fits neatly with many of our preconceptions. Yet the fossil and climatological data on which Wegener relied were suggestive but scarcely conclusive; paleontologists were themselves divided in their interpretations and conclusions. Moreover, if we attribute the skepticism of Wegener's ideas to his status as an outsider, what then should we make of the early treatment of Lawrence Morley? Harry Hess? Tuzo Wilson? Although they were established insiders within the field of geophysics, their ideas also met with harsh criticism, their papers with rejection. It is a truism in science that the papers that present the most radical—and perhaps ultimately the most important—scientific ideas of their day are often the ones that meet with the harshest reception during peer review. And, again, rightly so. Within any field of science, the body of collective wisdom is hard won, based on data and inference that have themselves survived the intense scrutiny of review. Scientists whose quest for truth upsets the apple cart will likely chafe at the resistance they encounter but will ultimately accept the responsibility to persevere.

In 1928, geologist Arthur Holmes proposed a model for mantle convection and plate motions that was not far off the mark. Applying the standards of scientific advancement to his own theories, Holmes wrote, "purely speculative ideas of this kind, specifically invented to match the requirements, can have no scientific value until they acquire support from independent evidence."[6] That is, testable hypotheses that future data can prove or disprove are essential before a hypothesis can pass from the realm of cocktail party conversation to good science.

What Wilson and Morley had that Wegener (and, indeed, Holmes) lacked was not acceptance in the club—or even a willingness to persevere—but rather the good fortune to have lived at the right time. Had Wegener not agreed to the expedition that claimed his life in 1930, he would have been in his early seventies when the great bounty of bathymetric data was collected in the early 1950s. Those who portray Wegener as a victim of the scientific establishment sometimes gloss over the fact that Wegener became part of the academic establishment in the 1920s, when he accepted a professorship at the University of Graz in Austria. It requires no great stretch of the imagination to suppose that, as a tenacious and energetic individual with the security of a tenured academic appointment, Wegener might have remained active in science well into his seventies. He might have been well positioned to be among the first to recognize the significance of the bathymetric data and to capitalize on it.

But he wasn't. And it wasn't a matter of fault, his or anybody else's. It was sheer happenstance that Wegener died just as Harry Hess, Lawrence Morley, and Tuzo Wilson emerged on the scientific scene. In politics, one person can perhaps a revolution make, but if and only if world events have first set the stage. In science the stage itself is more important than the intellect, personality, vision, or charisma of any single individual. The birthright of scientists is the state of understanding at the time that they take their places in their chosen field. The legacy of any scientist is the additional contributions he or she provides. The advancement of scientific understanding is a little like the building of a transcontinental railroad. The work to be done at any given time depends on the lay of the land. Even with tremendous scientific insight and acumen, it's hard to build a bridge if the tracks haven't reached the river yet.

TWO SIZING UP EARTHQUAKES

> The line of disturbing force followed the Coast Range some seventy miles west of Visalia and thence out on the Colorado Desert. This line was marked by a fracture of the earth's surface, continuing in one uniform direction for a distance of two hundred miles. The fracture presented an appearance as if the earth had been bisected and the parts had slipped upon each other.
>
> **–FROM THE *HISTORY OF TULARE COUNTY*,** on the 1857 Fort Tejon,
> *California, earthquake*

On a balmy autumn evening not so long ago, a woman was working in her yard when her two-year-old son figured out how to turn on the sprinklers. For the boy, turning a spigot and watching the aquatic fireworks was no doubt wonderful magic. Mom, however, was not amused. She admonished her child and returned to work. After one too many repeat transgressions, she scooped up the little boy, deposited him in his crib for a time-out, and returned to the front yard to finish her work.

Within minutes the ground began to tremble with a vengeance well beyond her previous experience. Hearing a terrified shriek from inside the house, the woman rushed back inside, her heart pounding, to discover that her child was mercifully unscathed but clearly scared out of his wits. The boy remained quiet and somber until his father arrived home from work not long thereafter. With wide-eyed earnestness, the little boy implored his father, "Daddy, don't turn on the sprinkler."

The date was October 17, 1989, and the setting was the San Francisco Bay Area. The earthquake was the M6.9 event that would come to be named for the mountain peak that was the most striking geographic feature in the immediate vicinity of the epicenter: Loma Prieta. Its lilting name notwithstanding, any hint of poetry associated with this earthquake was belied by its societal impact: sixty-three dead, $6.8 billion in direct property damage, and as much as $10 billion in total damage including indirect losses.

In the weeks and months following the earthquake, seismologists and geologists worked to understand the enigmatic event. It had occurred near the San Andreas Fault but was atypical of San Andreas events of its magnitude for two reasons: the quake had neither the pure lateral (strike-slip) mechanism nor a rupture that broke through to the surface. Had it generated fault rupture that reached the earth's surface but was difficult to observe in the mountainous terrain? Had it occurred on the San Andreas Fault or on a separate thrust fault immediately adjacent to the San Andreas?

A certain little boy, however, knew from the beginning what fault was involved: his own. Not long ago, intelligent and educated adults had no better understanding of earthquakes—their basic mechanics and causes—than did that little boy. And although small or distant temblors might seem manageable enough to be little more than curiosities, the sheer power and horrible violence of a large earthquake almost invariably leaves humans of all ages feeling humbled and deeply terrified.

Imagine awaking in a heartbeat from the oblivion of sleep to the sensation that your house has been plucked off its foundation by a giant who is shaking it back and forth in his fist. Through a thundering rumble from the earth itself, you perceive other noises: explosive shattering of glass inside and outside your home, heavy furniture toppling over, explosions from electrical transformers outside, power lines crackling loudly as they arc and then fall, your family members screaming from their rooms. You cover your head—getting out of bed, much less going anywhere, is physically out of the question—as a second distinct jolt hits within seconds, this one even stronger and longer in duration than the first. You lose all sense of time. Whether the shaking lasts for 10 seconds or 10 minutes, you know only that it feels like an eternity. A single thought repeats in your head: I am going to die.

Finally the shaking and the noise begin to abate. The roaring and crashing die down, and suddenly the single loudest noise you hear is the clamorous symphony of car alarms, every last one in your neighborhood, blaring frantically. You fumble for your bedside light, and it isn't there. It wouldn't matter if it were because the power is out, in your house and in the homes of several million of your neighbors. Possibly for the first time ever, the background glow of city lights is gone. You are enveloped by a darkness more isolating than any you have ever known though you dwell within the heart of a densely populated urban region.

Your children's continued screaming penetrates the swirling fog inside your head as the realization dawns: there is a sea of broken glass and toppled furniture, and heaven only knows what else, between you and them. In the next instant a sickening realization dawns: you never did get around to strapping down your hot water heater or installing that emergency gas-shutoff valve.

For those whose experience with temblors is direct rather than academic, an earthquake has nothing to do with slip on a fault but everything to do with the abrupt and sometimes violent effects that result from fault movement. It is little wonder that such earthshaking experiences have over the ages inspired interpretation in terms of divine retribution and wrath.

This chapter, however, will explore earthquakes from a seismologist's point of view, which means a close-up view of the earthquake at its source. What is an earthquake? How is earthquake size quantified? How are earthquakes recorded and investigated? The answers to such questions are the building blocks that must be laid in place before we can look at topics like earthquake prediction.

You might imagine the basics of earthquakes to be dry science—knowledge long since established, grown dusty with age. Ironically, though, some of the most ostensibly simple questions regarding earthquakes have proven to be among the toughest nuts to crack. We know that plate tectonics is the engine that propels the earthquake machine. The motion of the plates generates the stress, directly or indirectly, that ultimately drives faults to fail, abruptly in earthquakes. What is less clear is the nature of the transmission associated with this machine. Exactly how and when do faults fail? Why do earthquakes stop? How do ruptures travel along complicated fault zones? Although some aspects of seismology were indeed established long ago, a discourse on the subject leads quickly to issues that are far more complex than they seem, issues that remain the focus of active research. Any presentation of earthquake basics inevitably raises almost as many questions as it answers.

FAULTS

The conceptual model for earthquakes, known as *elastic rebound theory,* is younger than the equations that describe electromagnetic theory and not much older than Einstein's theory of relativity. Developed by geologist H. F. Reid af-

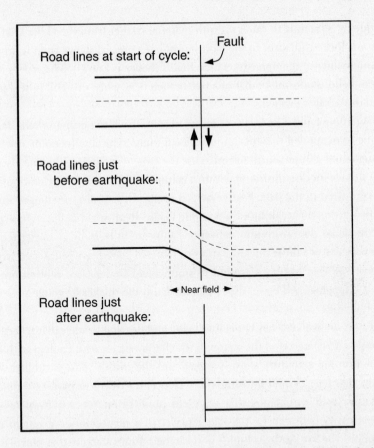

Figure 2.1. Effect of elastic rebound theory on road lines that cross a fault. A locked strike-slip fault will stay pinned right at the fault. Well away from the fault, however, the crust will continue to move. The curved road lines in the figure indicate a zone of elasticity in which the crust becomes strained to some degree. When an earthquake occurs, the fault and the broader strained region lurch forward to catch up. Over a full earthquake cycle, the entire crust on one side of the fault is moved with respect to the crust on other.

ter the great San Francisco earthquake of 1906, the idea behind elastic rebound theory is illustrated in Figure 2.1. A volume of crustal material near a locked fault is subjected to slow but steady bending, or strain, as indicated by the curved road lines in the figure. Friction results in the accumulation of stress in the vicinity of the fault (in what geologists call the *near field*). When the stress becomes sufficiently high to overcome friction, the crust adjacent to the fault suddenly rebounds to catch up with material farther from the fault. That the crust is locked adjacent to the fault but not at a great distance from it implies elasticity, hence the name by which Reid's theory is known. The crust's ability to bend is not unlike that of a tree branch bent when one end is held fixed and the other is pushed.

What is a fault? The origin of the word traces back to mining, where "fault" is the term applied to a break in rock continuity, typically the point at which water enters a mine. Earth scientists use the same term to describe a nearly planar surface (or discontinuity) within a volume of rock, along which movement has occurred in the past. Faults—which occur in the midst of stronger, intact crust—represent weak links in a strong chain. Real-world faults are not perfectly planar discontinuities, however. Faults are universally characterized by the presence of a finite thickness of worn material known as a *gouge zone*, which is material that has been ground to smithereens by repeated motion on the fault. The older and better developed the fault, the thicker its gouge zone will typically be.

Faults do evolve. They come into being and develop in ways that reflect the stresses and character of the region in which they are located. Fault growth and evolution are extremely slow, however, and the earth's crust is riddled with faults that have existed for some time. Hence, for the most part, earthquakes are associated with faults that represent pre-existing zones of weakness in the crust. A fault can be the major player in a plate-boundary system (for example, the San Andreas Fault), or it can be a minor fracture that moves only infrequently.

Earthquakes happen because the forces of plate tectonics (and sometimes other forces, such as postglacial rebound, whereby the crust rebounds upward after the retreat of large ice sheets) provide a source of stress. A stress, in a scientific sense, is simply an applied force over a given area. In an elastic medium, stress produces warping, or strain, such as that represented by the curved road lines in Figure 2.1. The technical meanings of these terms are not too different from their colloquial meanings: the traffic jam that makes you late for work is the stress; the pounding headache that the traffic produces is the strain.

A small minority of faults move steadily and slowly (or "creep," in the vernacular) in response to the stress applied; such faults never store enough energy to generate more than a steady crackling of small to moderate earthquakes. In more typical situations however, faults are locked, unable to accommodate motion except by way of the abrupt jerk of an earthquake rupture. The nature of the locking mechanism has been an active topic of research over the past decade or so. One question seismologists have asked is Why do earthquake ruptures come to a stop once the locking is overcome in a catastrophic failure? According to current theories, ruptures may stop because they encounter parts of the fault that have not stored enough stress to sustain a rupture; ruptures may also stop where they encounter geometrical complexity in the fault.

Geometrical complexities may play a role in causing faults to be locked in the first place, although the more important mechanism is thought to be friction. This mechanism seems obvious enough: we would expect there to be friction between adjacent blocks of rock, especially at midcrustal depths (that is, at depths greater than a few kilometers), where the pressure exerted by the overlying crust (the *lithostatic pressure*) is enormous. With such stresses, everything is densely packed. Open fractures in the crust are squeezed shut; even weak earth materials, such as fault gouge, become dense and therefore strong.

If, however, friction at depth is necessary and sufficient to explain fault locking, how does one explain the existence of creeping faults? The answer to this question is not entirely clear, but creeping faults may involve the presence of lubricating fluids within some fault zones. In any case, creeping faults are, by far, a minority class; to an overwhelming degree, faults in all tectonic settings are generally locked.

In the 1980s, a new conceptual paradigm emerged to explain the rupture processes of locked faults: the asperity model. *Asperities* are small sections of a fault that are relatively strong. A long, locked fault is pinned at its asperities, unable to rupture over its full extent until these strong points are overcome by the driving stress.

The asperity model has conceptual appeal, but it results in the same paradox: How can any segment or patch of a fault deep in the earth's crust be considered weak? This conundrum has led many Earth scientists to propose a conceptual model in which asperities are the geometrical complexities of a fault zone. Figure 2.2 shows a map of real faults in Southern California. Once again, the earth reveals itself to be far more complicated than simple cartoon representations. A structure that might appear at first blush to be a single, through-going fault invariably comprises a collection of smaller fault segments. Although certain segments might appear straight and continuous, closer magnification reveals finer-scale fragmentation—just as a branch, when viewed close up, resembles a tree in shape and complexity. In a mathematical sense, beasts with this property are said to be *fractal*, or self-similar. Perhaps, then,

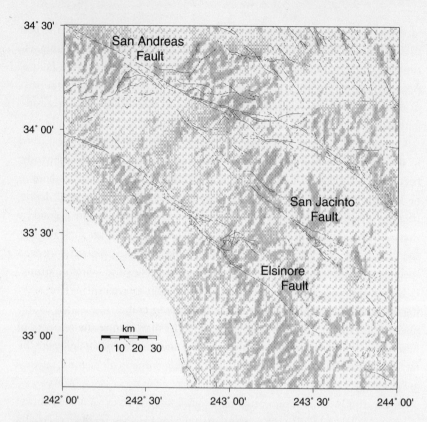

Figure 2.2. Fault systems in Southern California, including the San Andreas, San Jacinto, and Elsinore. Each of these complex fault systems comprises innumerous smaller segments.

it is fragmentation—the small kinks and offsets—in a fault that gives rise to asperities.

In the final analysis, we do not completely understand faults. Theoretical and computational models have been developed to simulate fault ruptures. Such efforts represented an intellectually lively and challenging specialty of geophysics through the 1990s and will likely continue to do so for some time. The first class of such models to be developed, known as spring and block models, attempts to characterize faults with mathematical equations that describe blocks connected to one another by one set of springs and coupled to a resisting force by another set (Figure 2.3). The first set of springs serves as a proxy for interaction between individual blocks, or elements, of crust; the latter set

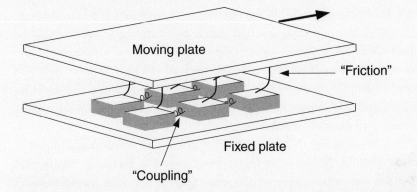

Figure 2.3. Spring and block model for earthquakes. The behavior of faults is modeled by a series of blocks connected to one another and to the fixed plate by springs. The springs between the blocks serve as a proxy for the interactions between fault segments (the "coupling"); the other set serves as a proxy for fault friction. If one block is pushed with enough force to overcome the frictional resistance, it will slip and push on its neighbors. If that additional push is enough to overcome the resistance of neighboring blocks, then the earthquake rupture will propagate. By constructing these models and introducing friction that varies among the blocks, seismologists can investigate the nature of earthquake ruptures on complicated faults.

represents fault friction. As a driving force is applied to the system, the blocks remain stationary until the frictional resistance of the weakest block is overcome. Once one block slips, it applies an additional stress to neighboring blocks, which will in turn fail if their frictional resistance is overcome. The blocks themselves, or sometimes groups of blocks, can be thought of as discrete fault segments with varying levels of friction.

Such simple models can reproduce characteristics of real-world earthquake distributions, which I present in detail in the next chapter. Numerical fault models developed in the 1980s and 1990s can also mimic the observed character of earthquake sequences. More-recent models consider the fault as a continuous structure rather than a series of discrete blocks; in addition, there are models that take into account the frictional and geometrical interactions among the innumerable particles of fault gouge. (Seismologists who develop models such as these typically excel at theory and at burning through substantial computer processing time on the most high-powered computers available.)

When is a model right? When can we conclude that a series of mathematical equations—inevitably a simplification no matter how sophisticated the theory or the computers—captures the essence of fault and earthquake behavior in the real world? Such questions are the stuff of which extensive and often heated scientific debate is made. With instrumental seismology and elastic rebound theory both barely a century old, the dynamics of earthquake rupture are not yet fully understood.

In a situation like this, it is important to stop and acknowledge imperfect understanding. To gloss over fundamental gaps is to present an explanation that will not—that cannot—make sense. Such an approach inevitably leaves astute readers scratching their heads and wondering what they have missed to feel so confused about such basic questions as, well, Why *do* earthquakes start? How *do* they stop? The "dumbest" questions people ask—the confusions they are left with—are often not dumb at all. If the experts cannot give you an answer that makes sense, it may be because none exists.

Nevertheless, to proceed with a look at earthquake basics, we must gloss over certain issues and move on. In a glossy nutshell, then, earthquakes happen along pre-existing zones of weakness known as faults. Plate tectonic and other forces provide the source of stress. Stress leads to strain accumulation because most faults are unable to accommodate the stress by means of steady creep. When the stress overwhelms the capacity of a fault to resist it, slip occurs along the fault in a sudden event we call an earthquake. Earthquake rupture might stop for one of several reasons: because the rupture encounters a patch of fault that has ruptured too recently in the past to have stored sufficient stress, or because the rupture encounters some sort of asperity and stalls in its tracks.

EARTHQUAKE SIZE

As I mentioned earlier, to most people an earthquake is less about a process on a fault than it is about the effects that fault rupture causes. The earth quakes. To seismologists, however, the shaking caused by earthquakes is known as ground motion. Although this topic is of considerable interest (see Chapter 4), ground motion is not an earthquake. To most seismologists, an earthquake is what happens on a fault, not the resultant shaking.

Earthquakes result from abrupt slip on faults; we know that. However, the overwhelming majority of events occur at least a few kilometers beneath the surface, well away from our direct scrutiny. We can find out about these earth-quake ruptures largely by virtue of the seismic waves they radiate. The energy that goes into seismic waves typically constitutes only a small percentage of the energy associated with earthquakes, the lion's share of which is expended to drive the motion across the fault. The leftover energy that is radiated provides the societal terror that earthquakes inspire and, more benevolently, bestows on us one of the best tools we have to probe the nature of the beast.

Before delving into the nature of seismic radiation and the ways that it elu-cidates physical processes associated with earthquake rupture, I would like to reiterate a fundamental characteristic of earthquakes: they represent the slip of an extended fault surface, not a process that occurs at a point. Although earth-quake rupture initiates at a single point that can be identified rather accurately, rupture propagates over an area that reflects the eventual size of the event. So, too, does the amount of motion (slip) on the fault scale with an earthquake's size. For big earthquakes, the large extent of fault rupture and slip can result in dramatic effects, as evidenced by the account in the chapter epigraph, writ-ten a half century before the development of elastic rebound theory.

Earthquake size. The concept might seem straightforward, but the quanti-fication of an earthquake's size is far from simple. How is a big earthquake dif-ferent from a small earthquake? How does one best describe the quantity "earthquake size"?

Although the public is most familiar with "magnitude" as a measure of earthquake size, that measure is a somewhat artificial construct, designed to be manageable and conceptually intuitive. The magnitude of an earthquake is not a physical quantity but rather a value that is assigned according to prescribed empirical rules. Originally developed by Caltech seismologist Charles Richter in the 1930s as a way to classify the relative sizes of earthquakes in California, the Richter magnitude scale is not a measure of a simple physical quantity such as temperature, length, velocity, or energy. Consider the units in which mag-nitudes are reported: not kilowatt-hours or Joules or any of their brethren with which physicists quantify energy or work. Richter simply chose "magnitude units." The Richter magnitude of an earthquake is determined from the max-imum deflection recorded on a particular type of seismometer and is corrected to account for the distance from the seismometer to the earthquake. Charles Richter devised an equation that took this value as an input and returned a user-friendly output: a value of zero for the smallest events that Richter be-lieved could be recorded on seismometers; a value of 3 to describe the smallest earthquakes that were felt; and values of 5–7 to describe the largest earthquakes that Southern Californians might experience in their lifetime (Sidebar 2.1).

Other magnitude scales have been developed that are based on peak seismometer deflections and the duration over which shaking exceeds a certain value (both rupture area and the duration of shaking relate to the dimension of the rupture, or source size). All such magnitude scales have their limitations. As classically defined, the Richter magnitude scale hits the stops, or saturates, at magnitudes larger than approximately 5.5. Above this value, calculated Richter magnitudes fail to rise proportionately as the earthquake size increases. The same problem plagues other scales based on peak ground motion amplitudes. Richter magnitude also cannot distinguish between earthquakes that generate the same peak amplitudes over longer or shorter durations.

Sidebar 2.1   The Test of Time

It is to be expected that technological advances will render some scientific conclusions obsolete. Physicists have their atom smashers, with which they have imaged successively smaller building blocks of matter. Astronomers use improving telescope technology to see progressively farther into the heavens. With more-advanced seismometer designs, Earth scientists have imaged earthquake sources with greater clarity and detail. But Charles Richter's 1930s definition of M = 0 has stood the test of time. Smaller events do occur, and they have been recorded on specialized, dense arrays of instruments, particularly arrays that include seismometers buried deep within boreholes. But even today, events smaller than M = 0 are virtually never recorded by modern, regional seismic networks. It is an impressive technological insight that successfully stands such a lengthy test of time.

To understand the quantification of earthquakes in the manner now employed by seismologists—a manner that reflects a true physical property of earthquake sources—we must revisit concepts that may be little more than distant, and possibly unpleasant, high school memories. On close examination, however, the concepts are far less alien than they seem, and far less difficult.

For a gross description of earthquake size and character, seismologists use two terms: *seismic moment* and *stress drop.* In physics, the moment of a force is nothing more than torque—the quantity that describes the turning capacity of an automobile engine. Torque has the same physical units as work; but the term "work" is generally used in the context of work done by a force, and "torque" generally describes the ability of a force to get work done. Understanding either concept inside and out requires equations only a mathematician could love, because one must account not only for the amplitude of the forces but also for their direction and the direction of the work done.

A full understanding of the concept of seismic moment as used by seismologists requires similarly serious mathematics: a driving force has a particular direction in three-dimensional space and will drive motion on a two-dimensional fault that has its own orientation. Fortunately, such complications are superfluous to an understanding of *scalar moment,* the overall amplitude of the seismic moment that is used to quantify earthquake size. An earthquake involves the interaction of the blocks of crust on opposite sides of a fault, a mechanical process that can be characterized by a torque, or seismic moment. Scalar moment can be shown to be equal to the multiplicative product of three terms: the area of the rupture, the slip produced by the earthquake, and a measure of material strength known as shear resistance (Figure 2.4).

Reporting earthquake size in terms of moment is inconvenient. Because of the enormous variation in earthquake sizes, observed earthquakes span a moment range of perhaps a dozen orders of magnitude. That is, if the smallest felt earthquake were normalized to have a moment of 1, the largest earthquakes that occur would have a moment of 1,000,000,000 (give or take). Although seismic moment best characterizes the size of an earthquake, the quantity is unwieldy, especially for use outside the scientific community.

To combine form and function, in 1979 seismologists Tom Hanks and Hiroo Kanamori introduced a magnitude scale based on seismic moment. Their moment-magnitude scale was designed to dovetail seamlessly with older magnitude scales for small to moderate earthquakes. To collapse a huge range of numbers down to a much smaller one, scientists often resort to the mathematical construct known as the logarithm (or log), another concept that is often rendered incomprehensible (and therefore unpleasant) in high school. Taking a logarithm involves nothing more than calculating the number of factors

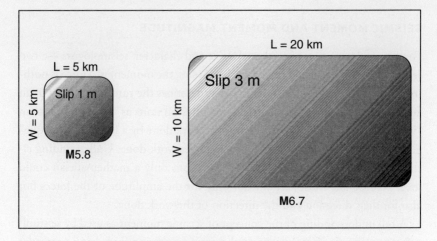

Figure 2.4. The relationship between fault area and moment magnitude. To obtain seismic moment from fault area, three terms are multiplied: the fault area (L × W), the slip, and the shear resistance. The seismic moment can then be used to calculate the moment magnitude.

of ten that a given number contains. That is, the log of 100 is 2 because 10 times 10 is 100; the log of 1,000,000,000 is 9 (10 × 10 × 10 × 10 . . . well, you get the idea). For a number that is not an even power of ten, the log is intermediate between that of the next lower and the next higher even powers. Thus, the log of 500 is greater than 2 and less than 3, intermediate between the log of 100 and that of 1,000.

Seismologists employ several techniques to calculate the seismic moment, including ones that rely on inferences made from recorded seismic waves as well as from direct field observation of rupture area and average slip. (Shear resistance is generally assumed to be fairly constant throughout the crust, perhaps varying as a function of depth.) Moment magnitude (M_w) is then calculated from seismic moment with a simple mathematical relation developed by Hanks and Kanamori. By definition, moment magnitude is consistent with Richter magnitude over a magnitude range of approximately 2–5. At higher magnitudes, M_w more faithfully reflects the actual size of an earthquake.

Although straightforward, the concepts involved with seismic moment and moment magnitude are not amenable to a twenty-five-words-or-fewer explanation. Confused by decades of brief and necessarily incomplete explanations from seismologists—who have developed at least a half dozen magnitude scales en route from Richter to Hanks and Kanamori—the newspaper editors of the world seem to have resorted at times to issuing clear marching orders to their troops. "Report the Richter magnitude, please, and skip the mathematical mumbo jumbo." Seismologists, meanwhile, know full well that the Richter magnitude of a large earthquake is almost meaningless. The solution is sometimes an imperfect but workable détente: seismologists report the magnitude they know to be best without making a big issue out of the scale used to derive it, and that value is sometimes erroneously described as being the Richter magnitude. Increasingly, though, the values are simply reported in the popular media as *magnitude*, without qualification.

The evolution of wisdom within the seismological community has injected confusion with regard to the magnitude of certain notorious historical earthquakes. Publications as recent as the 1980s list magnitudes of 8.2–8.3 for both the 1906 San Francisco earthquake and the great Fort Tejon earthquake that occurred on the southern San Andreas Fault in 1857. More-recent assessments of moment magnitudes for these events, assessments based on documented rupture lengths and slips combined with our current understanding of the depth of the San Andreas Fault, range from 7.7 to 7.9.

No primer on earthquake magnitude would be complete without a few comments on the subject of uncertainties. Few issues in seismology frustrate the public and the seismologist alike as much as the vagaries of magnitude estimation do. Even though agreement on the best methods for quantifying earthquake size has been established within the seismological community, issues related to magnitude uncertainty still arise. As recently as 1990, the weekly magazine of the *Los Angeles Times* ran a cover story on the 1989 Loma Prieta earthquake that was titled "The M7.1 Experiment." By the time the issue was published, the scientific community had settled on the moment magnitude, 6.9, as the preferred magnitude for this event. Given the proclivities of the human psyche, some small differences are more important than others. "Demoting" an earthquake from M7+ to M6+ is, particularly in the minds of those who experienced the horror of the earthquake, akin to denigrating personal suffering. Magnitude estimates that bounce around like rubber balls also have a nasty way of conveying a less-than-favorable impression of the collective competence of seismologists.

Three factors account for the lion's share of confusion regarding magnitudes. First, multiple magnitude scales do remain in use within the seismological community. Second, the growing need to deliver earthquake information in real time often forces seismologists to rely on a immediate result from automated analysis even though such rapid-fire estimates are often improved slightly by a more careful, human inspection of the data. Finally, for smaller earthquakes especially, magnitudes are often associated with real uncertainties of at least 0.1 magnitude unit either way. Therefore, the difference between M7.1 and 6.9 is both consequential and, with good data, resolvable, but the difference between M3.1 and M3.2 is probably not.

IS SIZE ALL THAT MATTERS?

Recall my earlier mention of a second quantity, stress drop, with which seismologists quantify the nature of the earthquake source. Fundamentally, stress drop is the simplest quantity seismologists have to describe differences in earthquake ruptures, differences that are critical for the nature of the ground motions generated by an earthquake. A vexingly elusive quantity to measure with any precision, static stress drop is the difference between the post-earthquake stress level on a fault and the pre-earthquake stress level. A bit of mathematical manipulation reveals that the stress drop of an earthquake boils down to the ratio of the average slip to its rupture length (or radius, if a rupture is approximately circular). The fact that this ratio seems not to vary in any systematic way over an enormous range of earthquake magnitudes is evidence that the physical processes governing earthquake ruptures do not depend on earthquake size. But the ratio of slip to length can vary by at least an order of magnitude from one earthquake to another. Such differences are scarcely inconsequential. For a given magnitude, an earthquake with a higher-than-average stress drop will likely be perceived as sharp. The ground motions from a high-stress-drop event might be higher than average in a way that has significant consequences for structures subjected to the shaking. People who live with earthquakes—for instance those who have experienced a sizeable mainshock and its aftershock sequence—often come to recognize differences in the character of events of comparable magnitude. Although many of these differences result from the observer's proximity to the source and the nature of the local geology, variations in stress drop lead to real variability in earthquake shaking.

Not all seismologists embrace the concept of stress drop with equal affection. For one thing, the uncertainties associated with magnitude estimates pale in comparison with the uncertainties associated with stress-drop estimation.

Given small to moderate earthquakes, for which neither the slip nor the rupture length is easy to infer, even the best stress-drop estimates are resolved within a factor of ten. Moreover, as seismologists have developed sophisticated theories and computer programs to reconstruct detailed patterns of slip in moderate to large earthquakes, the limitations of using a single stress-drop parameter to characterize a source have become clear. Even though one can estimate average slip and hence average stress drop, inferred slip distributions of large earthquakes reveal substantial complexity that cannot be captured by a single parameter.

In light of the limitations—both observational and conceptual—associated with stress drop, we are left with seismic moment and moment magnitude as the key parameters for describing the size of an earthquake rupture. Size might not be all that matters, but size is what we talk about because size is what we can measure.

ANATOMY OF AN EARTHQUAKE RUPTURE

Although the general physical processes that control rupture are thought to be the same for both large and small earthquakes, large events (M6.5+) are set apart from their smaller brethren in two critical respects. First, the former usually rupture the entire width of the brittle (seismogenic) crust, so their geometries are more rectangular than round. (In the seismological vernacular, "width" refers, perhaps perversely, to the vertical extent of faulting. "Length," then, is the horizontal dimension of a rupture.) Second, seismologists and geologists can investigate large events in substantially greater detail than can be done with small earthquakes. Even for a M4 event, the distribution of slip can be complex (Figure 2.5). Yet the limited extent of the source size limits the possible level of complexity; the limited source size also renders inscrutable any complexity. The larger an earthquake, the more there is to see.

Any earthquake, big or small, initiates at a point. The exact nature of the initiation process was the subject of considerable debate through the 1990s. Some lines of observational and theoretical evidence suggest that a gradual acceleration of slip—a slow nucleation process—precedes the abrupt onset of earthquake rupture (Figure 2.6). This nucleation might play out over a timescale of minutes or hours, or even days. But because such a gradual process would not generate seismic waves, it would remain invisible to standard seismic instrumentation.

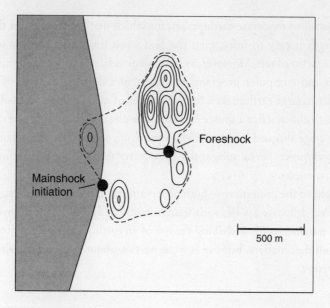

Figure 2.5. Rupture model for a M4.4 foreshock of the M6.1 Joshua Tree earthquake (1992). The model, derived by James Mori,[7] reflects the complexity of the rupture process even for this relatively small event. Here the contour lines correspond to different levels of slip within the overall inferred rupture dimension *(dashed line)*.

As far as people and seismometers are concerned, an earthquake begins at a point known as its *hypocenter,* in the instant that rupture begins and radiates seismic waves. (The more familiar term, *epicenter,* refers to the point on the earth's surface above the hypocenter.) In the mid-1990s, seismologists Greg Beroza and Bill Ellsworth presented evidence suggesting that large earthquakes start off looking like small earthquakes. Focusing on high-quality recordings of large events in California, these researchers showed that the waves from many large earthquakes are preceded (typically by a few seconds) by seismic waves that are indistinguishable from those produced by smaller earthquakes. The most straightforward interpretation of the Beroza and Ellsworth hypothesis is that large earthquakes are nothing but small earthquakes that get out of hand. According to this interpretation, a M7 event is the culmination of a cascade in which smaller events successively trigger larger ones.

Other models and evidence support an alternative hypothesis, which holds that the size of an earthquake is somehow pre-ordained from the first moment

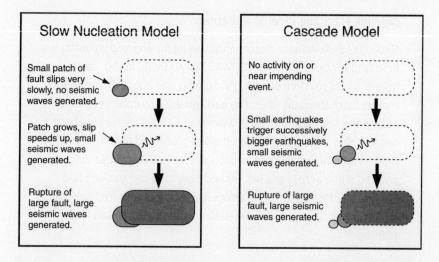

Figure 2.6. Preslip, or slow nucleation model *(left)*, and cascade model *(right)* for earthquake nucleation processes. In the former, an underlying, unseen process finally builds to produce low-amplitude shaking that marks the slow onset of a large earthquake. The large earthquake takes off in a breakaway phase characterized by strong shaking. In the cascade model, there is no underlying process, but only a series of small, closely spaced patches of slip that eventually trigger the large earthquake rupture. Adapted from work by Douglas Dodge, Gregory Beroza, and William Ellsworth.[8]

of initiation, either by the nature of the initial rupture or perhaps by the state of stress on the fault. In this interpretation, the slow start of large earthquakes documented by Beroza and Ellsworth could represent the effects—the modest snaps and crackles—associated with an underlying mechanism that drives the initial rupture process.

If the reason why large earthquakes happen as they do is unclear, how they proceed is much better understood. Rupture propagates away from the hypocenter, sometimes in both horizontal directions (bilaterally) along the fault but more commonly in one direction only. In 1990, seismologist Tom Heaton presented a model that brought our understanding of large earthquake ruptures into sharper focus. Based on inferred-slip models for large earthquakes and on compelling eyewitness accounts of earthquake rupture, Heaton's slip-pulse model proposed that the total slip at any point along a fault is achieved remarkably quickly, almost in the blink of an eye (Sidebar 2.2). At any given time during a large earthquake, slip occurs over only a narrow band along the fault. This slip pulse propagates forward as the earthquake proceeds. This process is analogous to pulling a large, heavy carpet across a floor by introducing a small wave, or ruck, that propagates the entire length of the carpet. Eventually the entire carpet will be translated from its original position, but at any one time only a small ripple is in motion.

An earthquake slip pulse can intensify in amplitude or weaken as it propagates; it can also slow down, speed up, or even jump to neighboring fault segments that are spatially noncontiguous but not far enough apart to halt the rupture. The rupture of faults across offsets has been investigated by seismologists Ruth Harris and Stephen Day, who used state-of-the-art computer modeling techniques. Here again, however, the reasons why these things happen are less clear than the result, which is that every large earthquake has its own unique blotchy pattern of final slip.

One of the important seismological developments of the 1980s and 1990s was improved methodology to image the details of large earthquake ruptures. Utilizing data from strong-motion instruments, seismologists have worked out analytical methods to interpret (or invert) recordings of an event and obtain a finite fault model that describes the blotchiness of the slip distribution. Like snowflakes, every earthquake is unique, its slip pattern controlled by innumerable details in the fault's structure and previous rupture history. The closer we look, the more detail we see—sort of the converse of Heisenberg's uncertainty principle.

PROBING THE SEISMIC SOURCE

In previous sections, I discussed salient features of earthquake ruptures and pointed out unresolved questions associated with them. I noted that earthquake scientists can investigate these issues because of the small percentage of energy associated with earthquake ruptures that goes into generating seismic waves in the earth. In this section I narrow in on the waves themselves.

When a fault ruptures, one side moves relative to the other. Because the movement is abrupt, it generates elastic waves in the surrounding crust. The simple push provided by each side of the fault generates compressional (or simply P) waves, the likes of which you can generate with a Slinky toy by stretching it taut along its long axis and then giving it a sharp push (Figure 2.7). Meanwhile, the opposing movement along the two sides of the fault generates a shear (or S) wave. Shear waves can also be generated in a Slinky with a sideways push that introduces a traveling S-shaped ripple.

Observation of P and S waves provides information about the earthquake rupture that produced them. Figure 2.8 illustrates a slipping patch of fault. Dividing the crust into four quadrants, we see that the fault rupture will compress the crust in two quadrants and extend it in the other two. If we look at the P waves in all four quadrants, we find that the first ground motions are upward (that is, outward) in the compressional quadrants and downward (inward) in the extensional quadrants. From this pattern of initial ground motion directions we can infer fault orientation, or what Earth scientists term the *focal mechanism,* although the inherent symmetry of the physics involved always leads to an ambiguity between two possible planes. To decide which plane to pick, we need more information. For example, the distribution of early aftershocks will usually delineate the direction of the fault's rupture.

P and S waves also elucidate an earthquake's location. Both P and S waves are generated the instant rupture begins, but they travel away from the source at different speeds. Although the physical processes that govern these waves are quite different from those governing lightening and thunder, the net effect is similar: the farther the two types of waves travel, the greater their temporal separation. The propagation times of the wave types can be codified into rules of thumb that relate the separation of the signals to their distance from their point of origin. With lightening and thunder, a 5-second delay corresponds to 1 mile; with P and S waves, each 1 second of separation corresponds to about 5 miles (8 kilometers) of distance.

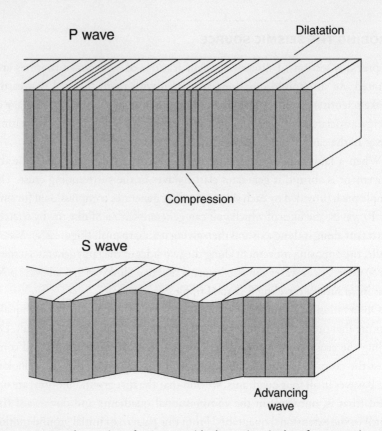

P wave

Dilatation

Compression

S wave

Advancing
wave

Figure 2.7. *Top*, the motion of a P wave, with alternating ripples of compression and extension; *bottom*, the motion of an S wave.

The easiest way to locate an earthquake is to observe the arrival times of P and S waves at a network of stations. The time difference, the S-P time, at any one station is enough to determine the distance from that station to the source; one cannot, however, determine the direction from which the waves originated. To determine location in three-dimensional space, observations from three or more stations are required. Figure 2.9 illustrates the method in two dimensions.

In practice, S-P times are rarely used in the routine analysis of network data. The theory behind this method is straightforward, but its implementation in the real world is fraught with complications. Most notably, although the initial arrival of the P wave can be identified ("picked") with some confidence because the wave emerges abruptly from the quiet background noise level, the

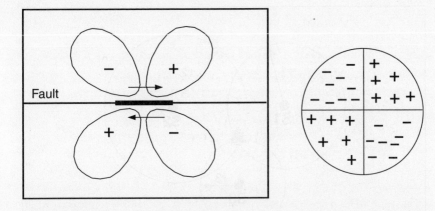

Figure 2.8. *Left*, map view of zones of compression and extension on a strike-slip fault. These zones can be used to infer fault orientation. Within zones of compression, the direction of the initial P-wave arrival will be up, or positive. In zones of extension, the direction will be down, or negative. *Right*, the focal mechanism. By observing the pattern of ups and downs in an earthquake, we can determine the focal mechanism. When represented in two dimensions, an earthquake focal mechanism always has two straight lines that divide regions of ups from regions of downs. One of these two lines corresponds to the orientation of the fault that ruptured during the earthquake.

precise arrival of the S wave is more difficult to pick because it arrives on the heels of the P wave. At a typical network site where data are recorded one hundred times per second, a P-wave arrival can be considered reliable when the analyst is confident that he or she can identify the exact sample point corresponding to the arrival. The precise S-wave arrival usually cannot be identified with the same precision.

Fortunately, location can be determined from P-wave arrival times alone. Although the conceptual ideas behind locating earthquakes by using P-wave arrivals are straightforward, in practice the process, like so many others, is surprisingly complex. A major complication arises from the fact that one is observing time but seeking to determine distance. If someone told you that he had driven for two hours to get to your house and then asked you to figure out how far he had driven, the task would be impossible unless you knew his speed en route. And if he told you that his speed had varied in a complicated way—for example, there may have been periods when he accelerated gradually—the task would be tractable but quite difficult.

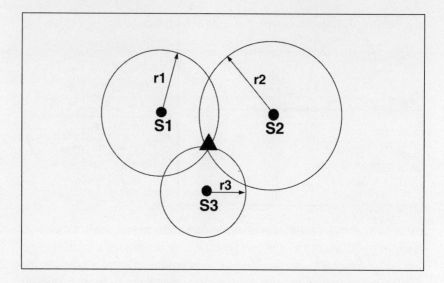

Figure 2.9. Earthquake location in two dimensions. The distances (r) between the earthquake source *(solid triangle)* and stations S1, S2, and S3 are determined from the arrival times of the P and S waves at each station.

In a similar way, to translate earthquake time measurements to distance, you need a model of seismic wave velocity in the crust. Seismic wave velocity, in turn, is a rather complicated beast, a function of rock type as well as depth in the earth. Determining an earthquake's location in the context of a complex velocity structure—and with data that have their own uncertainties—is mathematically involved. Precise earthquake location thus remains a topic of active research in seismology.

Methodology can determine "good enough" locations in a matter of minutes. Real-time computer software can monitor the output of a seismic network, identify events automatically, and determine a magnitude and a location accurate enough for most purposes. If we want to determine whether a certain cluster of earthquakes is spread out over 50 meters or 1 kilometer, then more sophisticated analytical methods are required (Sidebar 2.3). But if we want to know whether a hypocenter was in one town or a neighboring one, then routine methods will usually suffice. As computer technologies continue to improve, today's high-resolution methods will likely become the standard real-time analysis methods of tomorrow.

OBSERVING SEISMIC WAVES

Earthquake waves are recorded by instruments known as seismometers. The use of mechanical devices to record earthquakes dates back to China in the second century A.D., when Cheng Heng invented an elegant device consisting of a sphere with dragons, facing outward, affixed around its diameter. In the mouths of the dragons were placed balls, which were connected to an internal pendulum. The direction of ground motion was (supposedly) identified by observing which ball was dislodged.

Cheng Heng's ingenious invention would become no more than a footnote in the annals of seismology history, however. The first modern seismographs were not developed until the late 1800s. These early instruments employed a design still used by many recording devices more than a century later. Ground motions are recorded by observing the motion of a mass attached to a pendulum (Figure 2.10). Improved seismometer designs were developed in the late twentieth century. Paralleling late-twentieth-century advances in technology, there has been an ongoing and successful push to develop seismic instruments that are smaller, lighter, cheaper, and easier to operate. The scientific impetus to improve seismometer design is twofold: to faithfully record the entire spectrum of energy radiated by an earthquake and to faithfully record ground motions caused by earthquakes over a broad range of magnitudes.

To understand the first goal, consider a simple pendulum—a suspended mass free to sway back and forth. Such a pendulum will swing with a natural

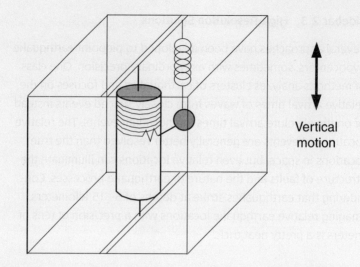

Figure 2.10. Simple inverted pendulum seismometer for recording vertical ground motion. Earthquake waves cause the weight on the spring to move up and down, and the motion of the weight is then recorded on paper. Modern seismometers are generally digital, however: their output is written to a computer file. For the most part, paper (drum) recordings—such as those one might see in the movies—have become obsolete.

frequency that depends on its length. If the frequency of ground motion co-incides with the natural frequency of the pendulum, then the pendulum will respond to the motion of the ground. If, however, the ground moves back and forth more slowly (that is, at a lower frequency), the pendulum will not reflect the motion of the ground.

Unlike pendulums, large earthquakes radiate energy spanning a sizeable frequency range, perhaps from 0.01 to 100 hertz (or cycles per second); that is, each or complete cycle (or period, the reciprocal of frequency) can last from 100 to 0.01 seconds. Earthquake ground motions are analogous to music in that they contain a range of tones, low-frequency (long-period) tones and high-frequency (short-period) tones. Building a seismometer that responds to high-frequency energy is generally easier than building one that can record lower-frequency energy. Relying on the simple pendulum design, low-frequency instruments quickly become unwieldy (Sidebar 2.4). Better designs allow low-frequency (long-period) seismometers to be built inside manageable packages.

Sidebar 2.4   A Matter of Physics

One of the counterintuitive tenets of basic physics is that the period of a simple pendulum depends only on its length, not its mass. (Hence the fact that children and adults on neighboring swings will swing at the same rate.) Specifically, the period is 2π times the square root of the ratio of pendulum length to the acceleration of gravity. The acceleration due to gravity is approximately 980 centimeters/second/second, so a 5-centimeter pendulum will have a natural period of approximately 0.5 seconds. Building a simple pendulum seismometer with a period of 10 seconds requires a pendulum length of 2,700 centimeters, or approximately 90 feet.

The second main goal of instrumentation development—to increase the range of magnitudes that an instrument can record on scale—has also led to big strides in the later decades of the twentieth century. The classic approach to covering the full range of earthquake magnitudes required different types of instruments: very sensitive weak-motion seismometers to record low-amplitude motions and insensitive strong-motion instruments for recording the largest earthquakes.

Because older weak-motion seismometers remain on scale only for the smaller earthquakes they record, they give travel time and first-motion information but not much else. That is, one can use data from these instruments to measure with some precision the arrival time of the P wave and its initial direction, but little else because the recordings are frequently off-scale. The primary focus of research with data from weak-motion networks, then, has been earthquake detection and the determination of their locations and focal mechanisms. Precise magnitude determination, especially for events bigger than approximately M3, has been difficult with data from classic weak-motion networks.

Strong-motion instruments provide data only from larger events but are designed to remain on scale over the full duration of larger earthquakes. These instruments, known as force-balance accelerometers, rely on a different design than do their weak-motion counterparts.

The two types of instrumentation led to the development of distinct subfields of seismology. Weak-motion seismology is concerned with earthquake locations and focal mechanisms derived from weak-motion data; these are generally useful for quantifying earthquake distributions and for understanding how an area is shaped by earthquakes. Strong-motion seismology, by contrast, is concerned with analyzing peak amplitudes and the entire seismogram (waveform), both of which elucidate the earthquake sources and seismic wave propagation. For many years, seismologists aligned themselves with one camp or the other, each with its own concerns and its own culture (Sidebar 2.5).

Sidebar 2.5 Culture Clash

The relationship between weak- and strong-motion seismologists has been respectful over the years. After all, they are working toward a common end: an understanding of earthquakes and the ground motions they produce. But scientific subfields do develop their own cultures, and sometimes those cultures clash. One clash concerns the portable seismic recording systems that were first developed and deployed by strong-motion seismologists. Such systems were often of little use to weak-motion seismologists because of the persistent problem of timing. That is, without a permanent, high-quality clock, the timing of portable instruments was too imprecise to yield useful arrival times, and therefore the data could not be used to determine earthquake locations. Over time, some people in the weak-motion community felt that because strong-motion seismologists did not need precise time information, developers of portable instruments were not improving the timing as fast as they might have otherwise. Technological advancements once again rode to the rescue with the advent of accurate timing using GPS satellites (see Chapter 1), and harmony now (generally) prevails.

In the 1980s and 1990s, the development of broadband seismic instruments—instruments that record ground motions over increasing ranges of amplitude and frequency—has heralded an exciting new era for observational seismology. For example, seismologists can now use the abundant recordings of smaller events to study the nature of the earthquake source and the propagation of ground motion, instead of being limited to the sparser data from large earthquakes. Seismology is a data-driven science—significant advances in instrumentation cannot help but move the field forward by leaps and bounds. The distinction of weak-motion seismologist and strong-motion seismologist is also, happily, receding into the past as new instrumentation unifies a once fragmented scientific community and a fragmented science.

Considering the history of observational seismology, one cannot help but recall the old tale about the investigation of an elephant by blindfolded observers who are each allowed to feel only a certain part of the animal. Seismologists are not only blindfolded with respect to any direct view of earthquakes, we are also allowed only brief glimpses of our study subjects. Earthquakes do not stand around like elephants. They are fleeting, transient phenomena that we will catch only if we are ready for them. From this perspective, it is perhaps less surprising that questions remain unanswered. Yet as seismologists endeavor to integrate their investigations and to develop better tools to compensate for their lack of direct vision, the future cannot help but seem bright. It will be no surprise if future advances in earthquake source theory come on the wings of major strides in instrumentation or data collection, or both. It has happened before; it will almost certainly happen again.

THREE EARTHQUAKE INTERACTIONS

An earthquake of consequence is never an isolated event.

—CHARLES RICHTER, *Elementary Seismology*

Charles Richter's observation, penned in 1957, would have been old news to Eliza Bryan almost a century and a half earlier. Little is known about Miss Bryan except that she lived in the tiny frontier settlement of New Madrid, Missouri, in 1811 and that she chronicled her own experiences of earthquakes of consequence. Bryan described a violent shock at about two o'clock in the morning on December 16, 1811, and the tumultuous cascade of events that followed. At daybreak on the same day, a large aftershock struck, nearly rivaling the mainshock in severity. Then, after a series of aftershocks, another powerful event happened on the morning of January 23, 1812. According to Bryan, this event was "as violent as the severest of the former ones." Bryan went on to write, "From this time on until the 4th of February, the earth was in continual agitation, visibly waving as a gentle sea." After a handful of strong shocks on February 4 and 5, the culminating blow struck during the wee hours of the morning on February 7, a "concussion . . . so much more violent than those preceding it that it is dominated the hard shock."[10] The December and January events caused substantial damage over a swath along the Mississippi River some 50 kilometers wide and 100–150 kilometers long; Eliza Bryan's observation that shaking from the earlier events paled in comparison to that from the February "hard shock" is sobering testimony indeed.

That earthquakes perturb the crust in such as way as to effect the occurrence of other events is no surprise to anyone who, like Eliza Bryan, has personally experienced an earthquake of consequence. In 1999, the lesson was brought home in a tragic manner to residents of Taiwan and northwest Turkey, battered first by substantial mainshocks and then, over the ensuing months, by one or more destructive aftershocks. (In Turkey, the August M7.4

Izmit earthquake was followed three months later by the M7.2 Duzce earthquake.) Scientists have known for some time that several mechanisms play a role in aftershock triggering, but a full understanding of the mechanisms behind earthquake sequences has proven elusive. It is not difficult to think of reasons why one earthquake should affect subsequent events; any earthquake of consequence will cause nontrivial alteration of the crust around a fault, including, probably, groundwater flow. Such changes provide a source of stress in the surrounding crust, stress that in some cases will trigger subsequent earthquakes.

The processes associated with earthquake triggering are complex, but significant progress in understanding them has been made within the span of a single decade. Indeed, the 1990s may come to be regarded as nothing short of a revolutionary time in our understanding of earthquake sequences, a time that gave rise to an entire new subfield of seismology concerned with earthquake interactions.

Development of earthquake interaction theories has been fueled by myriad factors, including a better understanding of earthquake ruptures themselves. Before going into earthquake interactions, I will step back and summarize some of the salient earthquake source results. Studies of the detailed, finite—fault slip models mentioned in the last chapter became ubiquitous in the seismological literature starting in the 1980s. The blotchy character of fault slip soon became familiar to Earth scientists. An earthquake of magnitude 6 or larger usually comprises subevents—isolated blobs of high slip within a broad region of lower slip. Imagine a long line of dominoes, all spaced closely enough to just barely brush their neighbors when they fall. Once a single domino begins a cascade, one might expect the entire grid to fall in sequence. But if some dominoes wobble but remain standing, then entire patches of the line may remain upright within the larger cascade. The final pattern will be a motley one, with an uneven pattern of standing and toppled blocks.

The final rupture pattern of an earthquake is inevitably mottled in a similar way; patches of slip are both lower and higher than average. Seismologists know the high-slip blobs as asperities, although it is not clear whether these blobs necessarily correspond to the frictional and geometric asperities mentioned in Chapter 2. (The latter were first introduced with the term "barriers," but this usage has largely given way to ambiguous usage of the single term "asperities.") The variation, or gradient, of slip in the vicinity of an asperity can be impressive.

A rupture model for the M6.7 1994 Northridge, California, earthquake is shown in Figure 3.1. Toward the edge of the largest slip concentration, a slip of approximately 4 meters is inferred. Scarcely 2 kilometers away, slip is approximately 1 meter. This difference might not sound problematic at first blush—2 kilometers is, after all, a significant distance. But consider taking a 2-kilometer-long freight train and forcing one end to move by 4 meters while constraining the other end to move by no more than 1 meter. How is the 3 meters of differential motion accommodated? If the cars are coupled with springs, the springs will compress or extend. But if both the cars and the couplings are rigid, the extra 3 meters could produce substantial strain along the train's length. If the train were rigid enough, something would have to give. A linkage component might break, or the train as a whole might derail.

To a good approximation, the earth's crust is rigid. Recall the basic tenets of plate tectonics theory: oceanic and continental plates experience extremely low levels of internal deformation over geological time. But, you might well be asking, what of the basic tenets of earthquake rupture theory? The elastic rebound theory discussed in Chapter 2? The two models are not strictly consistent. A perfectly rigid body would not have the elasticity to produce the curved road lines shown in Figure 2.1.

The solution to this apparent conundrum lies with two recognitions. The first is that a perfectly rigid body is a mathematical construct; all real materials deform under strain. The second is that the ability of the crust to bend elastically is quite small. For crustal rocks, a deformation of one part in ten thousand (that is, 1 meter of squeezing or stretching applied over 10 kilometers of crust) corresponds to a substantial strain. In the laboratory, samples of rock typically break at strains of a few parts in one thousand.

Let's return to the rupture model shown in Figure 3.1. The transition from the edge of the asperity to the lower-slip region corresponds to a strain of roughly three parts in two thousand (that is, 3 meters over the course of 2,000 meters). Clearly, then, the slip pattern as a whole results in a substantial and complex pattern of strain. And therein lies the basic physics—or at least part of the basic physics—behind Charles Richter's observation. An earthquake of consequence cannot happen in isolation because it produces too big a disturbance in the surrounding crust.

An additional element of the physics behind Richter's observation relates to a factor not generally considered until the 1990s: ground motions from a large

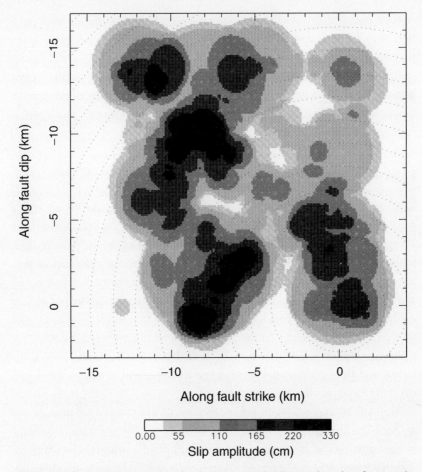

Figure 3.1. Slip amplitude contour for the M6.7 Northridge earthquake of 1994. The darkest shading indicates the regions of highest slip within the overall rupture dimensions. Adapted with permission from work by Yuehua Zeng and John Anderson.[11]

earthquake can significantly perturb the crust, sometimes at surprisingly long distances from a mainshock rupture.

Although facets of these realities were recognized and explored many decades ago, a mature and integrated theory of earthquake interactions did not take shape until the 1990s. These developments were driven by dramatic and incontrovertible observations from large earthquake sequences. In retrospect, some of these developments are so obvious that one might wonder why they weren't made sooner. And therein lies another tale of scientific revolutions—how they happen, and, perhaps more interestingly, why they happen when they do.

Before exploring the earthquake interaction revolution in detail, let me pause once again to first explain the classic, observational view of earthquake sequences. In this view, earthquakes generally belong to one of three categories: foreshock, mainshock, or aftershock.

EARTHQUAKE SEQUENCES: BASIC TAXONOMY

Earthquake taxonomy appears straightforward. An earthquake of any size is considered a foreshock if it is followed by a larger event in the immediate vicinity, typically no more than a few kilometers away and a few days later. As documented by seismologists Lucy Jones and Paul Reasenberg, only a small percentage of earthquakes will prove to be foreshocks. The existence of foreshocks and their documented characteristics raise interesting questions related to earthquake nucleation, which turns out to be relevant to earthquake prediction, as I demonstrate in Chapter 5. The statistics of foreshocks are relatively simple: in regions where data have been analyzed systematically (Italy, California, and Japan), about one in twenty small events will be followed within 3 days by a larger earthquake.

Although most small events are not foreshocks, about half of all mainshocks are preceded by one or more foreshocks. This asymmetry results from the magnitude distribution of events. The distribution of earthquake magnitudes within a region was documented many moons ago by Japanese seismologists Mishio Ishimoto and Kumiji Iida in 1939 and independently again in 1944 by Benito Gutenberg and Charles Richter. The Gutenberg-Richter (or b-value) relation describes the rate of occurrence of events with different magnitudes. The relation is given by the equation $\log(N) = a - bM$, or $N = 10^{(a - bM)}$, where N is the number of earthquakes per year of magnitude M or larger; a is a constant related to the overall seismicity rate in a region; and b is the slope of the line obtained when magnitude is plotted against the logarithm of the number of events (Figure 3.2). Regardless of tectonic setting, b-values close to 1 are observed almost universally. When $b = 1$, the a-value is simply the magnitude of the earthquake that occurs on average once a year in a region.

Given the mathematical form of the Gutenberg—Richter relation, a b-value of 1 can be translated into simple terms. Consider a region in which 1,000

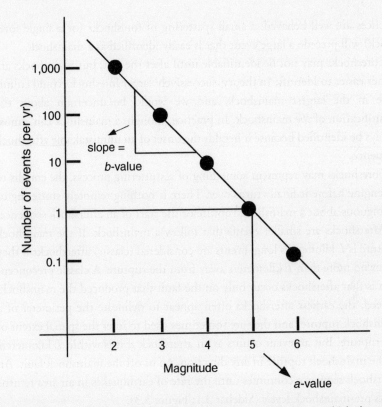

Figure 3.2. The *b*-value distribution of earthquake magnitudes, in which there are roughly ten times as many M5 events as M4 events, ten times as many M4 events as M3 events, and so on.

events above M2 occur in a year. According to the standard *b*-value distribution, roughly 900 of these events will be M2s (that is, M2.0–2.9), 90 will be M3s, 9 will be M4s, and perhaps one event will be a M5. If 1 out of 20 of the M3 events are foreshocks, then approximately 5 M3 foreshocks will occur in that year. But with only 9 events of M4 and larger, the foreshocks will precede a full half of all M4+ mainshocks. (This is a simplification: some of the larger events will themselves be foreshocks.) Foreshock observations can thus be summarized as follows: most (19 out of 20) earthquakes are not foreshocks, but about half of all mainshocks are preceded by foreshocks. A mainshock, meanwhile, is defined as the largest event that occurs within a sequence. Sometimes this simple taxonomy can break down, but for the most part earthquake sequences are well behaved: a small spattering of foreshocks (or a single fore-shock) will precede a large event that is easily identified as a mainshock.

Foreshocks may not be identifiable until after the fact, but mainshocks are rather easier to identify. In theory, successively larger foreshocks could culminate in the largest mainshock, and we would be uncertain about the identification of *the* mainshock. In practice, though, a mainshock can almost always be identified because it heralds the onset of an unmistakable aftershock sequence.

Foreshocks may represent something of a stuttering process, the coughs of an engine before it finally turns over. There is nothing remotely stuttering or ambiguous about a mainshock rupture or the start of an aftershock sequence.

Aftershocks are smaller events that follow a mainshock. If the mainshock rupture is L kilometers long, events are considered (classic) aftershocks if they occur no more than L kilometers away from the rupture. A classic preconception is that aftershocks occur only on the fault that produced the mainshock. Indeed, the earliest aftershocks often appear to delineate the perimeter of a mainshock rupture, and they are sometimes used to infer the spatial extent of the rupture. But an event counts as an aftershock if it is within L kilometers of the mainshock rupture in any direction, on or off the mainshock fault. An aftershock sequence continues until the rate of earthquakes in an area returns to its pre-mainshock levels (Sidebar 3.1; Figure 3.3).

Sidebar 3.1 Ghosts of Earthquakes Past

Although the rate of aftershocks decays as $1/t$ on average, the magnitude distribution of an aftershock sequence does not change with time. That is, the ratio of large to small events remains the same through a sequence. Large events are uncommon to begin with, and they become especially infrequent late in the sequence. Nevertheless "late large aftershocks" do occur, and they tax public sensibilities greatly. When a strong aftershock pops up apparently out of the blue, many months after the perceptible aftershocks seem to have ended, people are incredulous. "What do you mean, an aftershock?! Aren't we *done* with all that?" Seismologists have to remind themselves that their view of an aftershock sequence can be quite different from the public's.

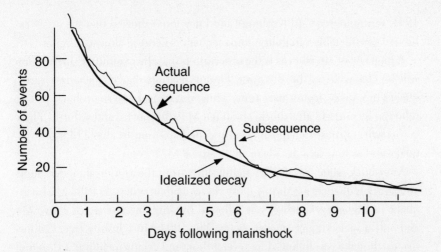

Figure 3.3. Rate of aftershocks as a function of time following a mainshock. The curve illustrates average decay. The rough curve illustrates the complexity of actual sequences, which do not follow the ideal model exactly but instead include subsequences as well as other complications.

As earthquakes go, aftershocks are well-behaved beasts. Although no individual aftershock is predictable per se, the magnitude, time, and spatial distribution of aftershocks generally adhere to established empirical laws. Empirical laws—laws, like the Gutenberg—Richter relation, derived from empirical data rather than from basic physical principles—are ubiquitous in the observational field of seismology; they are equations (such as the *b*-value relation) derived from repeated observation of the relationship between two parameters rather than from physical principles.

One of the oldest empirical laws of aftershock sequences, known as Bath's Law, holds that the largest aftershock in a sequence will generally be one magnitude unit smaller than the mainshock. In early 1990s, Lucy Jones examined modern data from earthquakes in California and refined this law. Although the largest aftershock does tend to be about one magnitude unit smaller than the mainshock, the largest event is equally likely to be anywhere from nearly zero to more than two units smaller than the mainshock.

A second law of aftershock distributions, established in 1961 by Japanese seismologist Tokuji Utsu and known as the modified Omori relation, describes the decay in aftershock rate with time after a mainshock. An aftershock sequence will decay, on average, with the reciprocal of time ($1/t$; Figure 3.3). In

1991, seismologists Carl Kisslinger and Lucy Jones showed that the decay exhibited considerable variability; some sequences decayed almost as fast as $1/t^2$.

A final law of aftershocks is the observation that the magnitude distribution will be characterized by the same *b*-value relation that characterizes earthquakes in a given region over time. Thus, the typical M6 mainshock will be followed by one M5 aftershock, about ten M4 aftershocks, and so forth. These events will continue to happen over time, perhaps months after a M5–6 earthquake and as long as a decade or more after a M7+ event.

Aftershock sequences do vary significantly from the canonical model. Some aftershock sequences are simply more energetic than others. A 1988 M6 earthquake in Saguenay, Quebec, was followed by only one aftershock above M4 and only a spattering of smaller events. The 1992 M6.1 Joshua Tree, California, earthquake was followed by several thousand events including a dozen of M4 or above. One should not, however, conclude that earthquakes in California produce more aftershocks than earthquakes in central or eastern North America. The 1811–1812 New Madrid earthquakes discussed later in this chapter produced thousands of aftershocks, possibly more than the great San Francisco earthquake of 1906.

Big or small, aftershock sequences establish their character early. The parameters describing the rate of aftershocks can be estimated within hours of a sizeable event. Although additional data will permit refinement of the early estimates, the data will rarely change the picture by much.

In 1989, seismologists Paul Reasenberg and Lucy Jones combined the established rules for aftershock sequences to derive an empirical relation for aftershock probabilities once a mainshock has occurred. Seismologists can calculate such probabilities immediately after a mainshock by using average sequence parameters tailored to fit the specific sequence. Seismic networks are now moving toward a routine broadcast of aftershock probabilities following significant mainshocks (Sidebar 3.2).

There is a phrase in the popular vernacular that makes seismologists cringe a little: "just an aftershock." After any felt earthquake, the press will invariably ask first about the magnitude and location and then follow immediately with "Was it an aftershock?" The "just" is implicit if not voiced: "If it's an aftershock, we don't have to worry about it." But aftershocks are earthquakes. Sometimes they are big, even damaging, events in their own right. And sometimes an aftershock will turn out to be part of an aftershock subsequence that pokes above the smoothly decaying curve that describes the rate. What is more, any aftershock can be followed by an event larger than itself, as Eliza Bryan learned back in 1812.

Sometimes activity in a region will die down after a mainshock, only to be ignited afresh by a second mainshock. Complications such as multiple, distinct mainshocks and aftershock subsequences begin to reveal the limitations of our simplest taxonomical categorization of earthquakes. Labeling events within a complicated sequence is sometimes a matter more of judgment than of hard science. Perhaps given the complexity of the crust and the significant perturbation a large earthquake causes, the surprise is not that earthquake sequences are so complicated but rather that they are often so simple.

THE 1992 LANDERS, CALIFORNIA, EARTHQUAKE SEQUENCE: NEW DATA, NEW INSIGHT

Seismologists' view of earthquake sequences was fundamentally altered in 1992, after the M7.3 Landers earthquake left a jagged tear through a sparsely populated part of the Mojave Desert on June 28. The largest earthquake to happen in California between 1952 and the end of the twentieth century, the

Landers mainshock was part of a remarkable sequence of events that illustrates nearly the full gamut of phenomena that are possible when an earthquake of consequence occurs (Figure 3.4). Observations from this sequence provided a bounty of data with which exciting theories could be tested; the quakes also inspired the development of wholly new theories of earthquake interactions.

To understand the Landers sequence, we have to step back 66 days to the first big earthquake in Southern California in 1992: the M6.1 Joshua Tree earthquake of April 23. The Joshua Tree sequence began on that day with a M4 foreshock and several smaller events. Although seismologists are aware that any moderate event might prove to be a foreshock, the M4 event on the afternoon of April 23 received more than its share of attention because it was within striking distance of the San Andreas Fault. As scientists reflected on the potential implications, a mainshock did occur several hours later—not on the San Andreas to the south but on a smaller fault to the north. Named for the national park the rupture bisected, the Joshua Tree earthquake was followed by an aftershock sequence that was as energetic (relative to its mainshock) as any in recent memory in California.

By late June the sequence was quieting down. Then on the evening of June 27 a tightly clustered sequence of events—twenty-two recorded by the Southern California Seismic Network—began pinging away near the small town of Landers, approximately 30 kilometers north of the Joshua Tree epicenter. The rumblings were frequent enough that at least a couple of Landers' residents decided to sleep in their cars. One person reportedly tried to contact someone associated with the seismic network but was unable do so because of the late hour. The largest of the twenty-two events was barely M3, not large enough to trigger the alarm set up to notify network analysts after hours.

Figure 3.4. *Bottom*, the locations of the 1992 Joshua Tree, Yucca Valley, Landers, and Big Bear earthquakes, as well as the surface rupture of the Landers earthquake *(dark lines)*; regional faults *(light lines;* faults are shown by solid lines where their surface expression is clear and by dashed lines where a continuous fault is inferred but difficult to trace on the surface); and the locations of M>5 aftershocks of the Landers earthquake *(dots)*. *Top,* the geometry of the fault jog in the Landers rupture. Because of the motion of the fault segments (illustrated by arrows), a small zone of extension, indicated by the dashed oval, is produced.

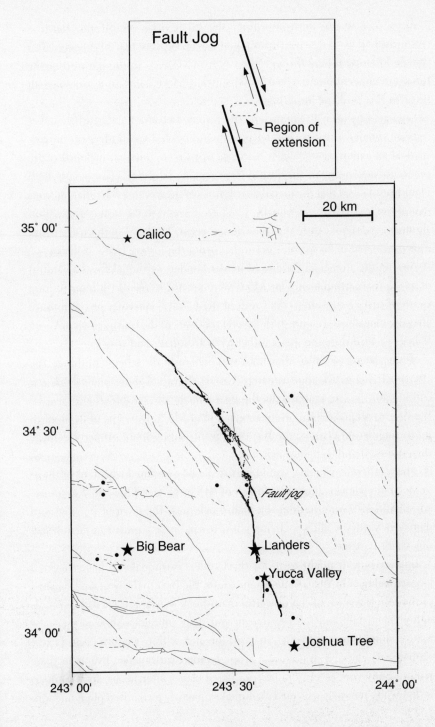

At 4:57 A.M. the next morning, the earth itself set off the alarm—enormously powerful waves from a mainshock rupture that ultimately left a scar 70 kilometers long through the California desert. (Although earthquake times are conventionally given in Greenwich Mean Time in seismology, all times in this book are local times.)

Separated by over 30 kilometers, the Joshua Tree and Landers earthquakes were considered distinct, but related, mainshocks. The Landers earthquake sparked an extensive aftershock sequence of its own, one that included many events to the south, in the Joshua Tree region. However, seismologist Egill Hauksson showed that the spatial distribution of these events was different from that of the Joshua Tree aftershocks. The rate of events in the Joshua Tree region also increased significantly after the Landers event. The later southern events are thus considered to be part of the Landers—not the Joshua Tree—sequence.

Among the aftershocks triggered by the Landers earthquake was a handful of substantial earthquakes. One M5.6 event created its own 11-kilometer-long swath of surface rupture to the south of the Landers epicenter just 3 minutes after the mainshock (Figure 3.4). Several residents of the nearby town of Yucca Valley reported thinking that another M7 earthquake had struck.

While the Yucca Valley aftershock was benign in the scheme of things, the aftershock that struck approximately 3 hours after the Landers mainshock was considerably less so. Dubbed the Big Bear aftershock—or sometimes even the Big Bear earthquake—this event weighed in at M6.5. By virtue of its proximity to population centers, the Big Bear event caused more property damage than did the Landers mainshock.

The Landers earthquake would eventually be associated with tens of thousands of aftershocks, including fifteen of M5 or greater. One M5.2 event occurred almost 9 months after the mainshock, near the town of Barstow, and damaged historic Calico, a "living ghost town," now a tourist attraction just outside of Barstow.

Impressive as it might seem for the Landers earthquake to have triggered large aftershocks in Yucca Valley to the south, Big Bear to the west, and Barstow to the east, the reach of this powerful mainshock was far greater still. Seismologists David Hill and John Anderson and their colleagues showed that seismicity along the eastern Sierran corridor and within western Nevada was significantly elevated in the wake of the Landers earthquake. Events in these remote areas, too far away to be considered classic aftershocks, soon came to be known as remotely triggered earthquakes, a newly recognized phenomenon.

These earthquakes generated significant interest among the seismological community. Later in the chapter, I delve into this issue at length.

Although by late 1999, the Landers sequence had apparently died down to a harmless trickle of aftershocks, an intriguing new chapter to the story was written in the wee hours of the morning on October 16 of that year. At 2:46 A.M., the M7.1 Hector Mine earthquake tore through the Mojave Desert, its rupture paralleling that of the earlier Landers mainshock. Separated by only 7 years and about 30 kilometers, the thought that the two earthquakes were not related defies imagination. Yet, as we will see shortly, existing theories of earthquake interactions do not explain how the first earthquake triggered the second. In any case, the Hector Mine event, considered a distinct mainshock, provides yet another illustration of the mystifying complexity of earthquake interactions.

THE NEW MADRID EARTHQUAKES OF 1811–1812

If this book appears to have a California bias, there is a reason. Of the twenty-three earthquakes of M6 or larger that occurred in the contiguous forty-eight United States between 1965 and 1995, nineteen were located in or immediately adjacent to California. (The remaining four were in three other western states: Idaho, Washington, and Wyoming.) Seismologists therefore look to the Golden State to investigate not only the specifics of California earthquakes and tectonics but also the general features of earthquake source processes. California by no means has a corner on the earthquake market, however. Seismicity rates in the other forty-seven contiguous states are simply lower, so we generally have to go farther back in time to find important events (see Chapters 7 and 8). In the central United States, we have to go back two centuries, give or take, to find to the most significant earthquake sequence in historic times: the great New Madrid sequence of 1811–1812, the effects of which were documented by Eliza Bryan and other early settlers. In 1811, New Madrid, which had been established as a Mississippi River trading post in 1783, was the second largest town in what is present-day Missouri, after Saint Louis. The town is approximately 165 kilometers north of Memphis, Tennessee, which was founded later but quickly eclipsed its ill-fated neighbor to the north. Although few people outside the Earth science community have heard of New Madrid, the dramatic seismic events of 1811–1812 are forever ensconced in the Earth science lexicon as the New Madrid sequence.

Because these events occurred well before the advent of instrumental seismology and at a time when the surrounding region was only sparsely populated, Earth scientists have had to be tremendously resourceful to glean information about the sequence. (Their endeavors are treated in detail in Chapter 7, in the context of paleoseismic investigations.) However, for the purposes of documenting the temporal progression of the New Madrid sequence, we can rely on contemporary accounts written by those who experienced the earthquakes firsthand.

Although inevitably subjective and imprecise, the contemporary accounts are consistent in documenting the salient features of the sequence. By all accounts, the sequence commenced with no warning—no foreshocks to foretell the drama that was to unfold on a cold and dark winter night. At approximately 2:15 A.M. on December 16, 1811, the earth trembled with a vengeance. The shaking—felt as far away as New York State and Ontario, Canada—caused minor damage as far away as Charleston, South Carolina. The magnitude of this event remains a source of debate to this day but has been estimated to be as low as M7.2 and as high as M8.1. The higher end of the range pegs the event as larger than either the great 1906 San Francisco earthquake or the 1857 Fort Tejon event in Southern California. Even if the true magnitude fell at the lower end of the range, the event would still have been a substantial earthquake, a repeat of which would have frightening implications for the present-day central United States.

Individuals living close to the New Madrid region described abundant aftershock activity in the hours following the first event. People as far away as western New York noted ground motions from a substantial aftershock that struck as morning dawned on December 16.

Between December 26, 1811, and January 23, 1812, observers described an Earth in almost ceaseless unrest. A remarkably careful resident of Louisville, Kentucky, by the name of Jared Brooks documented more than six hundred separate felt earthquakes in this time, and he cataloged the events according to perceived severity of shaking. Brooks categorized three events as Class 1 ("most tremendous," capable of causing damage) and four additional events as Class 2 ("less violent, but very severe").

On the morning of January 23, 1812, much of the eastern and central United States was again rocked by a large temblor, comparable to the December 16, 1811, mainshock. Successive aftershocks followed this event; Jared Brooks documented 209 events in the single week ending February 2, 1812.

But, as Eliza Bryan and others would learn, the earth had one last surprise up its sleeve. At approximately 3:45 A.M. on February 7, 1812, the earth trembled yet again, this time from an event that is widely agreed to be the largest earthquake of the entire sequence. Contemporary accounts note an increasing rate of felt earthquakes in the days leading up to February 7, quite possibly a foreshock sequence following very closely on the heels of the aftershock sequence from the January 23 event.

The February event came to be known as the "hard shock" among local residents. By virtue of its greater proximity to the town and, possibly, its larger magnitude, the February event reduced New Madrid to rubble. The earthquakes up until that point had damaged chimneys and caused significant slumping along the riverbank but had not shaken the houses hard enough to knock them down. Fortunately, by the time the houses fell, nobody was living in them anymore. The earlier temblors had been strong enough to persuade the residents of New Madrid either to flee or to leave their substantial houses for newly constructed, light wooden ones. Some months after the hard shock, the earth finally quieted down, although temblors continued to be felt throughout the decade.

Contemporary accounts of the New Madrid sequence are sobering. Even when not large enough to cause much damage, ground motions from large (M7+) earthquakes at even regional distances (100–1,000 kilometers) can be astonishingly powerful and frightening. Closer to New Madrid, the earthquakes were strong enough to disrupt the flow of the Mississippi River. An observer named Timothy Flint wrote, "A bursting of the earth just below the village of New Madrid arrested this mighty stream in its course, and caused a reflux of its waves, by which in a little time a great number of boats were swept by the ascending current into the bayou, carried out and left upon the dry earth."[12]

The extended nature of the New Madrid sequence must also have taken a heavy emotional toll, especially on residents who were close enough for the large events to be damaging and the small events perceptible. Imagine experiencing a major earthquake and, just as its extensive aftershock sequence finally begins to abate, another earthquake as large as the first hits. Now imagine the largest earthquake of all striking two weeks after that. It is unfortunate that so little data exist from the New Madrid sequence. Even two centuries later, though, it remains the most dramatic example of an extended earthquake sequence with multiple large mainshocks ever witnessed in the United States.

NEW THEORIES OF AFTERSHOCKS AND MAINSHOCKS

Because the behavior of aftershocks is straightforward, they were the first test-ing ground for theories of earthquake interactions. The $1/t$ decay rate of after-shocks led seismologists Amos Nur and John Booker to propose in 1972 that aftershock sequences were controlled by fluid diffusion. Diffusion usually refers to the physical process whereby a liquid or gas flows naturally into an-other liquid or gas or into a permeable solid. Mixing and migration are conse-quences of the random, thermally generated motion of molecules. Applied to aftershocks, the fluid-diffusion theory holds that a mainshock will disrupt groundwater, forcing fluid to flow away from regions where rocks are com-pressed and toward regions of extension, or dilation (see Chapter 2). In regions of dilation, the influx increases the pore pressure (the pressure of groundwater in rock pores), which reduces the frictional strength of faults. The weakening of faults in turn leads to aftershocks. A $1/t$ decay is a hallmark of a diffusive process—the effect of the disruption gradually diminishes as groundwater reestablishes equilibrium. Most of the re-equilibration is over quickly, but the system takes a long time to reach its final state of equilibrium ($1/t$ distributions have what scientists call "long tails").

Almost a decade after the Nur and Booker paper was published, seismolo-gists Shamita Das and Chris Scholz suggested another mechanism for after-shocks, one related not to fluids but to the stress and strain disturbances caused by a mainshock—an explanation related to the train analogy presented earlier in this chapter. Das and Scholz showed that a mainshock rupture would pro-duce a pattern of stress change with lobes of stress increasing at the ends of the rupture and, sometimes, off to the fault's sides as well. The pattern of stress change caused by an earthquake promotes earthquakes in some places and dis-courages them in other places. This mechanism accounted for the perplexing observation that some earthquakes created significant aftershock activity off to the sides of the fault that produced the mainshock.

The stress-triggering hypothesis proposed by Das and Scholz formalized the intuitive ideas of stress transfer by a large earthquake. Using a simple model for the crust, one can calculate the patterns of stress caused by a mainshock rupture. Such calculations are complicated because stress cannot be described simply by an amplitude; rather, it is associated with a directionality as well as with distinct shear and normal components. Shear stress is the stress that pushes a fault sideways; normal stress is the stress that acts perpendicularly on a fault, either compressing or decompressing it. The net effect of a stress change—whether it pushes a fault toward or away from failure—reflects a combination of both types of stress. Moreover, if you want to find out how a stress change with a given amplitude and orientation will affect the faults in a region, you must also know the orientation of those faults.

Because the crust is riddled with small faults of varying orientations, seismologists usually assume that a stress increase will create aftershocks on faults that are oriented properly for triggering. To test stress-triggering theory one usually calculates the pattern of stress change from a mainshock rupture and plots the maximum amplitude of stress at every point, in whatever direction the maximum happens to point. This pattern of stress change is then compared with the aftershock distribution.

Although the theoretical framework of stress-triggering theory was established in the early 1980s, extensive application and development of the methodology did not begin for almost decade afterward. Interest in the subject ignited like wildfire after the 1992 Landers sequence gave dramatic evidence of the significance of stress triggering. With data and a theoretical framework in hand, the setting was ripe for a major advancement in understanding earthquake interactions.

Within a couple of years of the Landers sequences, several journal articles were published that built upon the framework introduced by Das and Scholz. In 1994, Geoffrey King and his colleagues further developed the stress-triggering hypothesis and presented the implications of stress triggering for aftershocks and mainshocks. They pointed out that stress triggering could plausibly affect not only the occurrence of aftershocks but also the timing of subsequent mainshocks. That is, by raising stress in some areas, large earthquakes can cause subsequent large events to occur within months, years, or a few decades (Sidebar 3.3). Also in 1994, Jan Kagan showed that stress triggering could explain the tendency for large earthquakes worldwide to happen close in space and time. Ruth Harris and Robert Simpson called attention to the importance of "stress shadows"—regions where earthquake activity was predicted to be diminished because of the reduction in stress accompanying a large event in that region. (Earthquakes may raise the stress in some regions, but overall they relieve stress.)

In the years following the Landers earthquake, stress triggering became one of the hot topics in earthquake science. Scarcely did a major scientific conference go by without a special session devoted to the topic. If you want to find

Sidebar 3.3   Advancing the Clock

A fundamental tenet of earthquake theory holds that stress on a fault builds steadily over time as a result of plate tectonic and other forces. If a large earthquake increases or decreases static stress, that stress change will affect the timing of the next earthquake. Seismologists now speak of "clock advances" and "clock delays" to describe the effect on the timing of subsequent mainshocks. Unfortunately, earthquake interaction theory provides no answers to the critical question of where the clock was before it got nudged either forward or backward.

out what the exciting scientific developments in a field were at any time, you need look no farther than the abstracts of big conferences, in which summaries of oral and poster presentations are published.

In the Earth sciences, the fall meeting of the American Geophysical Union (AGU) has become *the* meeting to watch, the meeting scientists feel they must attend whether they want to or not. The abstract volumes for fall AGU meetings in the 1990s tell an interesting tale. In 1990 and 1991, papers on earthquake interactions were virtually non-existent. The 1992 volume includes perhaps a half dozen abstracts on earthquake interactions and stress triggering; all of them were related to the Landers earthquake and were presented in general sessions on earthquake rupture processes. A year later, there were a dozen presentations on the subject, ten in a special session devoted specifically to triggering. By 1995, a special session on triggering and stress transfer drew nearly two dozen talks covering earthquakes in regions as far-flung as Turkey, Japan, and East Afar. Later in the decade, the focus of the presentations broadened to address theoretical mechanisms of earthquake triggering as well as observational results. The observational studies became more sophisticated as well.

In 1997, seismologists Jishu Deng and Lynn Sykes presented an analysis of all large earthquakes in California going back to the Fort Tejon earthquake of 1857. By incorporating stress changes and the best estimate for the long-term stress rates due to plate tectonics, Deng and Sykes tracked the evolution of earthquake activity within the plate-boundary system. They showed that, to an impressive extent, large earthquakes influenced the location and timing of subsequent large events. Few earthquakes happened within stress shadows, and earthquakes occurred more often in regions where earlier events had increased stress.

Also in 1997, seismologists John Armbruster and Leonardo Seeber presented a new twist on triggering analysis. Instead of testing whether aftershocks of the Landers earthquakes had been triggered by the mainshock, Armbruster and Seeber assumed that they had been and then derived the distribution of mainshock slip that was most consistent with the aftershock observations. Their inferred model for the mainshock was strikingly consistent with models obtained from seismic data. Although other data with which the Landers source could be investigated were abundant, the Armbruster and Seeber results showed that aftershock distributions can be used to infer details of a mainshock slip pattern. These results suggest that for earthquakes that are not well recorded by instruments, obtaining rupture models might still be possible.

In August of 1999, the earth tragically verified stress-triggering theories when the M7.4 Izmit, Turkey, earthquake ruptured a segment of the North Anatolian Fault, which had been identified as ripe for future activity. Mindful of the notion that large earthquakes had marched from east to west across this major fault over the twentieth century (Figure 3.5), seismologist Ross Stein and his colleagues argued that stress had become significantly acute at both ends of the fault. In some ways, the progression of events on the North Anatolian Fault offers a textbook example of earthquake triggering. What is less clear is why no similar progression has been observed on the San Andreas Fault, which in many ways is quite similar to its distant cousin in Turkey. In any case, though, the orderly fall of dominoes on the North Anatolian Fault cannot help but raise serious concern for the future. Immediately to the west of the 1999 event, the next unruptured fault segment runs perilously close to Istanbul.

If the Izmit earthquake corroborates developing theories of earthquake interactions, the M7.1 Hector Mine earthquake of October 16, 1999, challenges them. Because the Hector Mine event was close to the 1992 Landers earthquake and parallel to the earlier rupture, its occurrence is difficult to explain. That is, the basic tenets of stress-change theory predict that Landers should have reduced the stress on the faults that ruptured in the 1999 event. The Hector Mine earthquake by no means invalidates the theories; instead it signifies a challenge that will likely herald a new level of understanding.

Theories of earthquake interactions will continue to develop and mature as scientists not only observe more sequences but also simplify assumptions and modeling approaches so that they more faithfully represent the behavior of the

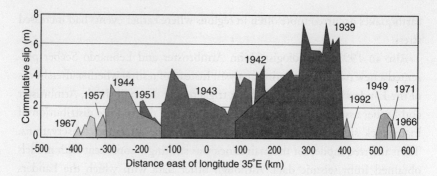

Figure 3.5. Progression of earthquakes on the North Anatolian Fault in northern Turkey. A large earthquake in 1939 increased stress at both ends of the rupture, apparently triggering a cascade of subsequent large events that played out over the rest of the twentieth century. Adapted from work by Ross Stein and others.

earth's crust. One modification, which was recognized early as important but difficult to model in detail, is the so-called viscoelastic relaxation of the subcrustal lithosphere. Recall that the brittle crust (where earthquake ruptures occur) extends only a few tens of kilometers in depth. Below the crust, the subcrustal lithosphere is more rigid than the mantle, but it accommodates deformation by way of plastic rather than brittle processes. The crust breaks; the subcrustal lithosphere oozes. Once a fault has ruptured in the brittle crust, the underlying lithosphere must somehow move (or relax) in response.

The brittle crust and the deeper lithosphere must be coupled; the shallow processes will influence the deeper ones and vice versa. The equations that describe the behavior of a three-dimensional system such as this are complicated and therefore difficult to model using computer programs. Only in 1999 did researchers Jishu Deng and Lynn Sykes (and others soon after) develop a program capable of realistic stress modeling of an elastic crust overlying a viscoelastic lithosphere. Deng and Sykes showed that slow ground deformations during the months following the Landers earthquake (the postseismic deformations) were most plausibly interpreted as a viscoelastic response. This process may have played a key role in triggering the 1999 Hector Mine earthquake.

The viscoelastic response of the subcrustal lithosphere is just one modification to the stress-triggering theory of Das and Scholz. At least three other factors must be accounted for before we can fully understand earthquake interactions: the role of fluids in the crust, the detailed nature of frictional processes on faults, and the effect of the dynamic strains associated with the passage of earthquake waves.

Although the focus of aftershock triggering moved away from the fluid-diffusion model of Nur and Booker, nobody doubts that fluids play an important role in aftershock triggering. That large earthquakes disrupt water within the crust is clear. Significant changes in spring and well levels following large earthquakes were documented at least as early as the 1811–1812 New Madrid sequence. But groundwater effects are difficult to model in detail because the crustal rock–fluid systems are extremely complex.

By investigating the aftershock sequence of the 1992 Landers earthquake, seismologist Greg Beroza concluded in 1998 that fluid pressure changes did play an lead role in aftershock triggering. Beroza examined the spatial variability of the decay rate of Landers aftershocks and found a slower decay in the extensional fault jogs. These jogs are offsets in a strike-slip fault system where a mainshock rupture will generally cause the region of the jog to either compress or extend, depending on the jog's geometry (see Figure 3.4). Beroza hypothesized that pore-pressure effects within the extensional jogs along the Landers rupture were particularly important because extension would have led to a fluid influx. Thus, whereas aftershock triggering in other areas may have been more strongly controlled by the instantaneous static stress changes, aftershocks within the fault jogs were controlled by longer-term mechanisms associated with fluid diffusion.

Also in 1998, William Bosl presented results from a very different type of investigation of fluid pressure and flow effects. Bosl used state-of-the-art computer algorithms to model the pore-pressure distribution caused by the Landers earthquake and compared his results to deformation and seismic data. The consistency between the predictions of his numerical simulations and the observational results lead him to conclude that pore fluid effects do play a role in the postseismic dynamic behavior of the crust.

Interest in fluid-diffusion mechanisms has persisted in part because stress-triggering theories—though conceptually attractive and apparently successful in predicting the spatial distribution of aftershock sequences—do not immediately explain one of the most salient features of aftershocks—their $1/t$ decay in rate. Bosl's analysis approaches an integrated model that incorporates stress-change and fluid-diffusion effects; however, such a model will inevitably be extraordinarily complicated and difficult to constrain in detail.

Another integrated theory emerged in the late 1990s that (apparently) can account for all of the established features of aftershock sequences. This theory—less conceptually intuitive than the fluid-diffusion theory—relies on the basic tenets of stress triggering, modified to account for the complex laws thought to govern fault friction. In 1994, Jim Dieterich posed the following question: If one started with a collection of faults in the vicinity of a mainshock, each governed by an appropriate friction law, how would the system as a whole respond to the abrupt stress change caused by the mainshock? Although the required calculations are mathematically involved, the results, in a nutshell, are that the model does predict many of the observed spatial *and* temporal distributions of aftershock sequences.

Although success in predicting observations is usually cause for celebration, aftershock sequences continue to present something of a conundrum. We know that several factors must conspire to control aftershock triggering, but we don't yet have a single unified model that includes all of these factors and shows how their confluence yields the characteristics of aftershock sequences that had been established by the 1960s.

In addition to static stress triggering and fluid diffusion, a final factor central to aftershock triggering is the dynamic stress change associated with earthquake waves. The stress change we have considered so far is what scientists term "static stress"—the fault slips and causes a fixed pattern of stress change that does not fade away with time. Consider, however, the sometimes sizeable seismic waves that ripple through the earth after a large earthquake. As the waves pass through the crust, they contribute a source of stress as rock experiences differential motion. This stress change is considered dynamic because it is transient; it lasts only as long as the ground is in motion.

After the 1992 Landers earthquake, when several seismologists were modeling the aftershock distribution with static stress change models, seismologist Joan Gomberg showed that the dynamic stress changes caused by seismic waves could have substantial amplitude within the aftershock zone. These results suggest that the dynamic stresses, which generally follow the same spatial pattern as the static stress changes, might also play a major role in controlling aftershock triggering.

The seismological community is left with what is almost an embarrassment of riches: too many apparently sound theories to explain a single set of observations. If all of the factors are important, with no one of them dominant—or, perhaps, with different ones dominant in different areas—then how do such simple aftershock sequence characteristics result? If one factor dominates the others, why are the other factors less important than it seems they should be? Such questions are vexing; they are also the stuff of which exciting research is made.

EARTHQUAKE TRIGGERING: BEYOND AFTERSHOCKS

I mentioned earlier that among the wealth of observations the Landers earthquake laid at Earth scientists' feet was the incontrovertible evidence that the earthquake triggered smaller events at far greater distances than the classic aftershock distribution would have predicted. Not fitting the mold of standard aftershocks, these triggered earthquakes merited a niche of their own in the seismological lexicon. Although hints of such a phenomenon had been floating around the seismological literature and lore for some time, the Landers earthquake was without question a watershed event. So dramatic and clear were the observations of earthquakes being triggered at distances greater than 400 kilometers that data demanded theory. The phenomenon was observed again after the M7.1 Hector Mine earthquake (Figure 3.6).

The scientific community responded with developments along two theoretical lines. First, Gomberg's dynamic stress results provided a plausible mechanism to account for distant triggering. By comparing observations of where triggering occurred in Landers with observations that triggering did not occur after other recent M6–7 events, she was able to quantify the amplitude and frequency range of ground motions capable of triggering distant events.

But there is more to the story. The Landers earthquake didn't simply trigger an indiscriminant spattering of events at regional distances. Instead, it preferentially—although not exclusively—triggered earthquakes in regions where high concentrations of fluids exist in the crust: geothermal regions such as The Geysers (north of San Francisco), which are associated with extensive groundwater systems, and volcanic regions such as the Long Valley caldera. This observation led immediately to theories describing the interaction of seismic waves with what scientists call "multiphase fluid systems" (systems involving liquid and gas). Curiously enough, several lines of evidence indicated that the humble bubble is critical in earthquake triggering. In 1994, Allan Linde showed that the disruption of bubbles in a deep fluid system at depth could create stress changes large enough to trigger earthquakes, a process known by the ungainly name of *advective overpressure*.

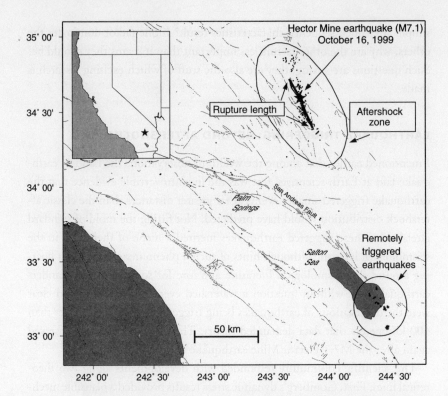

Figure 3.6. The aftershock sequence of the October 16, 1999, Hector Mine earthquake. As illustrated here, aftershocks are generally considered to be small earthquakes that occur no farther from a mainshock rupture than the rupture is long. Like the 1992 Landers earthquake, which struck in a similar location, the Hector Mine earthquake was followed by remotely triggered earthquakes at several locations, including the area near the Salton Sea. The small inset shows location of the Hector Mine epicenter within California.

Triggered earthquakes are, in a sense, easier to understand than classic aftershocks because one of the potentially important controlling factor—static stress change—can safely be dismissed in favor of dynamic stress. Simple physics reveals that the amplitude of static stress changes decays rapidly with distance (r) from a mainshock, as $1/r^3$. In contrast, dynamic stress associated with seismic waves decays more slowly, as $1/r$ or $1/r^2$, depending on the type of wave. So, oddly, our understanding of a phenomenon whose existence has been established only recently (remotely triggered earthquakes) is clearer than our understanding of a related phenomenon (aftershocks) whose characteristics have been known for many decades. Efforts to understand remotely triggered earthquakes are still in their infancy. These events are exciting to seismologists, however, because they stand to provide important new information about how and why earthquakes occur.

EARTHQUAKE INTERACTIONS: A REVOLUTION IN THE MAKING?

The 1990s witnessed tremendous advances in our understanding of earthquake sequences and earthquake interactions. Refinement of the most simple stress-triggering theory enhanced our understanding of aftershocks and helped scientists grasp the intricacies of large earthquake interactions. Further development of these theories continues apace.

Modern theories of fault friction appear to be important in earthquake triggering, and they continue to be explored. In 1999, seismologist Ruth Harris used Jim Dieterich's results to elucidate the nature of stress shadows. In particular, she wanted to know if the complex, sometimes counterintuitive, behavior of faults with complex laws of friction could account for the large earthquakes that occasionally pop up in regions where static stress modeling says they should not. If an earthquake decreases stress in a region, then no large earthquake should occur in that region until loading has added back that same amount of stress, and then some. While earthquakes generally seem to respect stress shadows, they sometimes do not. Such exceptions to the rule can undermine the validity of the theory as a whole.

Indeed, Harris showed that with complex friction, a large earthquake would push a fault toward or away from failure, but in a more complicated way than predicted by the simple stress-change theory. In particular, the closer a fault is to failure at the time that a mainshock on another fault happens, the smaller the stress change's effect will be. A stress decrease will still delay rupture of an event that was poised to happen soon, but sometimes by only a small amount of time.

The viscoelastic model of Deng and Sykes may also ultimately contribute new insight into the nature of large earthquake interactions, although such investigations are now only in their infancy.

Thus do the pieces of an integrated theory begin to fall into place. The development of earthquake-triggering theory in many ways parallels the development of plate tectonics theory. In both cases, data were collected for some time as glimmers of ideas took shape and led to early theories. Although both theories may seem obvious in retrospect, a point of synergy (for want of a better word) had to be reached, a point where abundant and high-quality data were obtained just as our conceptual and theoretical understanding had matured sufficiently for the data to be fully exploited.

The development of computer technology also played a nontrivial role in the development of earthquake interaction theories. Not only did better computer hardware and software allow scientists to analyze more data, in more complex ways than ever before, it also enabled us to display results in three dimensions and in color. Researchers thus could conceptualize complicated spatial processes much more easily than when they were limited to small data sets, two-dimensional images, and shades of gray (Sidebar 3.4).

The earthquake interaction revolution is much more an event in progress than the plate tectonics revolution that began at least two decades earlier. Much work remains to be done before we can fully understand Charles Richter's prescient observation: "An earthquake of consequence is never an isolated event."[13]

Much work remains, also, before we can exploit our burgeoning understanding of earthquake interactions. Although stress-triggering theories do not seem to be promising for short-term prediction, the results do suggest appli-

---

Sidebar 3.4   Pretty Pictures

The computer revolution has undoubtedly proven to be a boon to virtually all fields of earthquake science. Researchers can now analyze far more data, in far more sophisticated ways, than they ever could have dreamed of as recently as the 1970s (when a simple programmable calculator retailed for a few hundred dollars!). We can also display results in fancier ways. The explosion of color graphics in the 1980s and 1990s led some scientists to warn against "pretty picture syndrome"—the tendency to be uncritical of results that come in pretty packaging. Although it is indeed important that our attention not be shifted from the results themselves, differentiating results from packaging has become more difficult in recent years. In the analysis of complex three-dimensional data, presentation begets effective visualization. Visualization, in turn, begets understanding. In science, a pretty picture can be an important contribution.

Sidebar 3.5   Charles Richter

Those who knew Charles Richter often resort to the same small handful of words to describe him: brilliant, eccentric, and passionately interested in earthquakes. He was not known for his sense of humor, but glimpses of another side of the man can be found in the pages of his landmark book *Elementary Seismology*. One footnote on the issue of timekeeping tells the tale of astronomer who, in 1918, was stationed at an army post where a cannon was fired every day at noon. The astronomer asked how the base kept such precise time and was told that the officer in charge set his watch according to a clock in a jeweler's window. The astronomer then talked to the jeweler, who (you're ahead of me, right?) reported, "They fire a cannon at the fort every day at noon and I set my clock by that."[14]

cations for intermediate-term earthquake forecasting and, possibly, aftershock forecasting. If detailed mainshock rupture models can be determined quickly enough, regions at high risk for sizeable aftershocks (à la the Big Bear aftershock) could be identified in advance. Over the longer term, stress changes could help shape regional hazard assessment: hazard forecasts might increase in regions with significant positive stress changes and decrease in regions with negative stress changes. Such developments are not, I should emphasize, in our immediate future; the theoretical models and the observational results need time to mature before they can be applied to matters of societal concern. Within a few short months in 1999, major earthquakes gave us hope that stress-triggering theories might some day be exploited for hazard mitigation, and they also imbued us with a sense of caution about the pitfalls of applying theories that are still being developed. Yet the prospect of using earthquake interaction theories to improve earthquake forecasts is certainly not confined to the realm of science fiction. It appears to be on the horizon, within our grasp to reach some day.

Charles Richter had a long and extraordinary life (Sidebar 3.5). It is a shame that he missed the development of modern earthquake interaction theories. He would have enjoyed it.

FOUR GROUND MOTIONS

Science! true daughter of Old Time thou art!
Who alterest all things with thy peering eyes.
Why preyest thou thus upon the poet's heart,
Vulture, whose wings are dull realities?

—EDGAR ALLAN POE, *"To Science"*

When the Loma Prieta earthquake struck shortly after five o'clock on an October afternoon in 1989, the fault motion was restricted to a patch 15–20 kilometers long in the Santa Cruz Mountains, 80 kilometers south of the densely
populated San Francisco Bay Area. Yet within seconds of the event, the
double-deck Nimitz Freeway in Oakland was shaken hard enough for the upper deck to collapse onto the lower deck. Dozens of cars were trapped and
many lives were lost on that autumn afternoon. Only a stroke of remarkable
good fortune kept the death toll as low as it was—a World Series game was about
to commence in San Francisco, and many Bay Area commuters were already
home instead of on the road as usual.

Seismometers deployed in the aftermath of the temblor produced un-
ambiguous evidence that the freeway collapse was largely attributable to the fact
that it was built over mud, specifically the layer of bay mud that had been covered with artificial fill decades earlier so that the bay wetlands could be developed. Encountering the soupy mud, waves from the Loma Prieta earthquake
were slowed and amplified, which greatly enhanced the severity of shaking compared to the shaking at nearby sites on harder rock (Figure 4.1). A similar phenomenon led to substantial damage in the Marina district of San Francisco.

Sizing up potential earthquakes on known faults in a given region and determining their rates of occurrence are but the first steps toward an understanding of seismic hazard. Once source zones are identified and quantified,
one must then quantify the nature of the shaking that each event will gener-

ate. Earthquake prediction, earthquake interactions, plate tectonics—this is the stuff of sexy science, science that grabs the imagination and appeals to, rather than preys upon, the poet's heart. But although scientists and many non-specialists would agree that artistry and science are scarcely mutually exclusive, there has been little poetry to be found in the branch of seismology that addresses the question of earthquake shaking, or ground motions.

The seismological community has long since recognized the need to distill the artistry of earthquake science to a collection of prosaic but useful laws that can be used to predict the shaking caused by future earthquakes. Ground motion seismology is where the rubber meets the road in the translation of scientific knowledge to social issues such as building design and long-term hazard assessment. Forecasting the long-term rate of earthquakes doesn't forecast earthquake hazard unless one can calculate how future earthquakes will shake the ground, everywhere.

Two approaches have been employed to study ground motions. One approach is a classic, almost entirely empirical approach in which earthquake shaking is described rather than explained; the other is a more modern approach based on the physics of wave propagation. At present, seismic hazard maps are generated by means of the descriptive approach because the latter research yields results that cannot be readily incorporated into large-scale, automated analyses. Even so, in this chapter I focus on the physical approach—the state of the art in ground motions research—which represents the frontier of science that will ultimately shape the hazard maps of tomorrow.

The subfield of ground motions seismology traces its roots back to Japan in the 1870s, when the government instituted a policy of bringing in foreign experts, including physics and engineering professors. Several foreign and Japanese scientists endeavored to record earthquake waves with mechanical devices; the pace of the efforts accelerated after a large earthquake struck Japan in February of 1880. The first seismograms revealed a complexity that had not been envisioned by early theoreticians working on earthquake wave propagation. This complexity forced scientists to consider the question that has remained central to ground motions research to this day: How does one sort out the effects of the earthquake source from the complications introduced as the waves propagate?

The early investigations in Japan were focused on local earthquakes. Then, in 1889, a geophysicist by the name of Ernst von Rebeur-Paschwitz made an unexpected discovery as he measured ground tilt in an effort to investigate

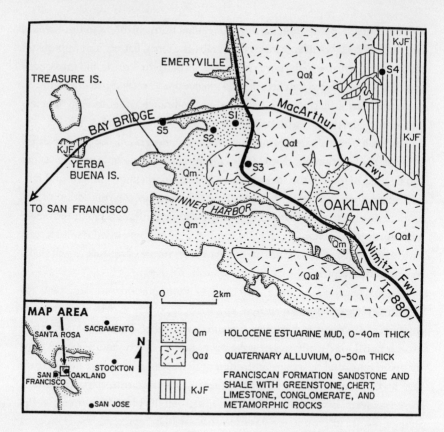

MAP AREA

SANTA ROSA   SACRAMENTO

N

STOCKTON

SAN FRANCISCO   OAKLAND

SAN JOSE

Qm   HOLOCENE ESTUARINE MUD, 0-40m THICK

Qal   QUATERNARY ALLUVIUM, 0-50m THICK

KJF   FRANCISCAN FORMATION SANDSTONE AND SHALE WITH GREENSTONE, CHERT, LIMESTONE, CONGLOMERATE, AND METAMORPHIC ROCKS

0   2km

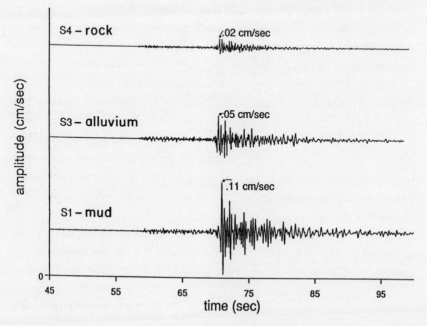

tides. Von Rebeur-Paschwitz noticed a signal on his instrument in Europe that he concluded had resulted from a large earthquake in Tokyo. Thus did scientists first realize that seismic waves from large earthquakes could be recorded by sensitive instruments halfway around the world. Some seven decades later, the great 1960 Chilean earthquake provided the first evidence that, in addition to generating the usual traveling waves, large earthquakes will "ring" the planet like a bell. Although the waves' amplitude is far too low to be perceived without highly specialized instruments, the so-called normal modes of the earth can last for hours, even days, after great earthquakes.

The study of earthquake waves thus encompasses global as well as local phenomena. Ground motions research, however, concentrates on seismic waves large enough to be felt—waves powerful enough to cause damage. Although seismologists often use recordings of small earthquakes to help understand wave propagation, the goal of ground motions research remains the understanding of ground shaking of societal consequence.

Almost as soon as seismologists began recording seismic waves, they realized that ground motions at close distances were extraordinarily complicated. Seismograms do not directly reflect the earthquake source; instead they reflect a complex combination of the source and propagation effects (Figure 4.2, *top*). Faced with such intractability, theoretically inclined seismologists for a long time zeroed in on global investigations (Figure 4.2, *bottom*), which were easier to understanding with the available data, theories, and computer resources.

Ground motions research largely became the purview of observational seismologists, who focused on the empirical rather than the theoretical. And thus was the poetry lost.

Figure 4.1. *Top,* map of the Oakland, California, area, including the Nimitz Freeway *(dark line)* and the Bay Bridge. The geology of the region is also shown, as described by the legend. During the Loma Prieta earthquake, the two-tier Nimitz Freeway collapsed over the extent of the segment that had been built on bay mud. Solid circles show the locations of seismometers deployed following the earthquake to record aftershocks. *Bottom,* an aftershock recording at three stations: S1, on mud near the freeway collapse; S3, on firmer sediments near a section of freeway that did not collapse; and S4, on harder rock in the Oakland hills. This aftershock occurred some 80 kilometers from Oakland. The dramatic difference in shaking at the three sites reflects the effects of near-surface geology. Illustrations by Kazuko Nagao.

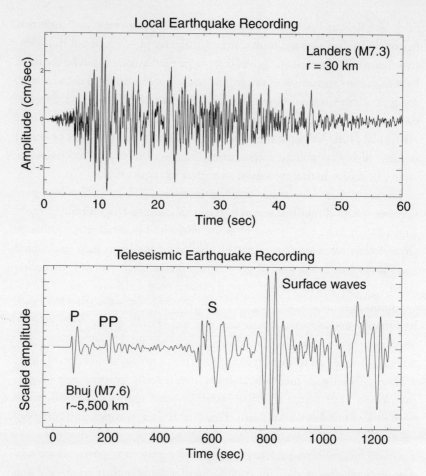

Local Earthquake Recording

Landers (M7.3)
r = 30 km

Amplitude (cm/sec)

Time (sec)

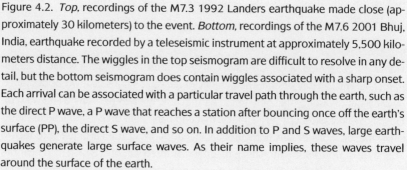

Teleseismic Earthquake Recording

Scaled amplitude

P PP S Surface waves

Bhuj (M7.6)
r~5,500 km

Time (sec)

Figure 4.2. *Top,* recordings of the M7.3 1992 Landers earthquake made close (approximately 30 kilometers) to the event. *Bottom,* recordings of the M7.6 2001 Bhuj, India, earthquake recorded by a teleseismic instrument at approximately 5,500 kilometers distance. The wiggles in the top seismogram are difficult to resolve in any detail, but the bottom seismogram does contain wiggles associated with a sharp onset. Each arrival can be associated with a particular travel path through the earth, such as the direct P wave, a P wave that reaches a station after bouncing once off the earth's surface (PP), the direct S wave, and so on. In addition to P and S waves, large earthquakes generate large surface waves. As their name implies, these waves travel around the surface of the earth.

POETRY FOUND: WAVES ARE AT THE SOURCE

If we want to move beyond an empirical characterization of ground motions, we need to know more about the underlying physical processes. To comprehend what happens to waves after they leave an earthquake rupture plane, we need to grasp their nature at the source itself. I introduced P and S waves in Chapter 2; however, a full characterization of seismic waves involves not only the wave type but also the nature of the radiated source spectrum. The spectrum of a seismic signal is simply a breakdown of how energy is distributed as a function of frequency (Figure 4.3). Spectra can also show the distribution of wavelengths. A child's prism generates the spectrum of sunlight, the range of

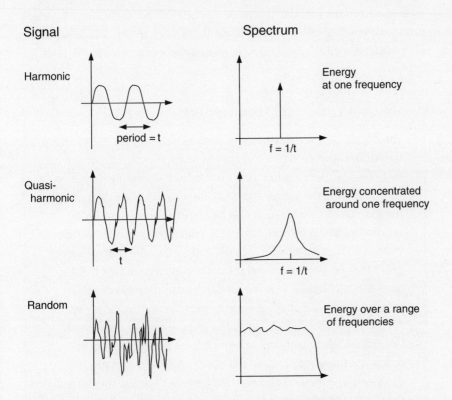

Figure 4.3. *Left,* earthquake signals: *top,* harmonic; *center,* quasi-harmonic; and *bottom,* random. *Right,* the spectra corresponding to the signals on the left: *top,* a spike at the one harmonic frequency; *center,* a smeared-out spike corresponding to the dominant frequency; and *bottom,* a range of frequencies up to some maximum value.

visible wavelengths corresponding to the familiar rainbow of hues. But whereas the spectrum of white light contains all colors (that is, all wavelengths) in equal measure, many other natural signals have more complex spectra. In a signal that contains a range of frequencies, certain frequencies—or certain ranges of frequencies—often dominate. In this sense, earthquake waves are not unlike music in that high tones and low tones are mixed together, although not randomly.

Efforts to understand earthquake spectra date back to nearly the earliest days of instrumental seismology. By the 1940s, investigators had observed that large earthquakes generate more long-period (low-frequency and thus low-energy) waves than do small earthquakes. In 1967, seismologist Keiiti Aki published a landmark paper describing source spectral shapes of previously published earthquake rupture models. Aki compared these models to data from recorded earthquakes and concluded that one model, known as the "omega-squared" (ω^2) model was most consistent with observations (Sidebar 4.1).

Sidebar 4.1   Lessons to Learn and Relearn

The classic papers in seismology, such as Keiiti Aki's 1967 source spectrum paper or the 1935 paper in which Richter introduced his magnitude scale, are often remarkable for their insight. Realizing that observing the source spectrum directly would be difficult, Aki ruminated over a technique that had been proposed earlier by seismologist H. Berkhemer in 1962. With this technique, the spectrum of a large earthquake was divided by the spectrum of a small event that occurred in a similar location. Because waves from both earthquakes experience similar propagation effects, this process yielded a corrected source spectrum for the large event. Although the theoretical contributions in Aki's paper went on to become widely known and referenced, the community did not recognize the significance of the observational insight Aki had proposed. The method of correcting large earthquake spectra with small event spectra was not developed until it was rediscovered— and dubbed the "empirical Green's function" method—many years later. The lesson for young scientists? Respect your elders, and read their papers.

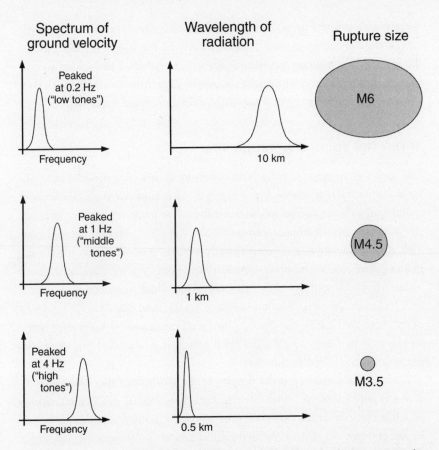

Figure 4.4. *Left,* the spectrum of earthquake shaking (ground velocity); *center,* the wavelength of radiation; *right,* the magnitude of the event. The dominant energy of a large earthquake is at low frequencies, analogous to low tones in music. Small earthquakes, however, predominantly radiate high-frequency energy, analogous to high tones.

We can perhaps grasp the ω^2 model most easily by looking at the ground velocity of seismic waves. The model predicts that a spectrum of ground velocity will peak at a certain frequency and then decay with the reciprocal of frequency on either side of the peak (Figure 4.4). The frequency at which the peak occurs is known as the *corner frequency.* In the simplest rupture model, this frequency is inversely related to the rupture's length. Because the energy of ground motions is a function of their velocity squared and because frequency is inversely proportional to wavelength, the interpretation of the spectrum shown in Figure 4.3 is straightforward: the dominant wavelength reflects the dimension of an earthquake's rupture. The bigger the earthquake, the longer the dominant wavelength of radiated energy. The high-frequency energy of a little earthquake might rattle your cage (literally and figuratively); the low-frequency energy of a big earthquake can topple buildings.

AFTER THE SOURCE: ATTENUATION

The spectrum of seismic waves is described by their source model—but only for a fleeting instant. As soon as waves leave a mainshock rupture, complex modifications begin. Waves spread out into an ever increasing volume of crust, just as ripples from a stone tossed into water will grow into ever larger circles. This effect alone, known as geometrical spreading, goes a long way toward explaining why ground motions decrease dramatically as they travel away from a rupture. In February of 2001, this effect was illustrated dramatically by a M6.8 earthquake near Seattle. Although larger than the devastating 1994 Northridge earthquake, the Pacific Northwest event was 50 kilometers deep. By the time its energy reached the earth's surface, geometrical spreading had significantly weakened its punch.

Geometrical spreading is the simpler of the two factors that affect seismic waves as they propagate. The second, more complicated one is attenuation, which is yet another seismological issue whose conceptual simplicity is belied by the enormity of effort expended to understand it. Attenuation is thornier than geometrical spreading for two reasons. First, geometrical spreading can be predicted from basic theoretical principles of wave propagation, but attenuation is controlled by multifaceted processes. Second, whereas geometrical spreading uniformly lowers the amplitudes of ground motions at all frequencies, attenuation affects some frequencies more than others—not unlike a stereo equalizer—and therefore alters the shape of the spectrum in ways that can be tricky to grasp.

That waves attenuate beyond the diminution caused by geometrical spreading is clear; exactly how and why they attenuate is murky. Attenuation mechanisms fall into two general classes: intrinsic and scattering. In the former, wave energy dampens by virtue of the inelastic nature of rocks. For example, as neighboring rock grains move, friction will absorb a small percentage of the energy associated with seismic waves. Attenuation of this sort is described by a quality factor, Q, defined such that the fractional loss of energy per wavelength of propagation is given by $2\pi/Q$. (Note that according to this definition, a higher Q corresponds to lower attenuation. The higher the Q, the more efficiently waves propagate.)

For a given source—receiver distance (that is, the distance between the source and the receiver), a high-frequency wave travels many more wavelengths (cycles) than does a low-frequency one, so attenuation typically affects high-frequency waves more strongly than it does low-frequency waves. Earthquake waves can thus have markedly different characteristics depending on the combination of their magnitude and the distance at which they are felt. At close distances the high-frequency source energy of a M3–4 earthquake will result in an abrupt but brief jolt. At large distances the high-frequency energy of a large earthquake will be attenuated, but the lower-frequency waves will still pack a wallop. At hundreds of kilometers from the epicenter, waves from a large (M7+) earthquake feel strong enough to wake the dead. But because their high-frequency energy is strongly attenuated, they usually pose only limited hazards to the living.

The second mechanism, scattering, contributes to attenuation and otherwise shapes the nature of ground motions. Unlike intrinsic absorption processes, scattering does not make energy go away but instead redistributes it. As waves hit boundaries between different types of rock, a fraction of the incident energy will be scattered, or diverted. We know scattering must be an important effect in the crust because seismograms show complex tails known as *codas*. That is, wave theories predict that, in a uniform medium, P and S seismic waves will arrive at distinct times (these waves are known as the direct arrivals; Figure 4.5A). In a layered crust, theory predicts a suite of distinct arrivals that correspond to the direct waves plus distinct reverberations from the layers (Figure 4.5B). But even compared to the ground motions predicted for a layered structure, actual seismograms are a mess (Figure 4.5D). Obviously the scattering must be complicated if it can shift energy away from the direct P and S waves to the prolonged codas we observe.

In 1973, having turned his attention from the seismic source to wave propagation, Keiiti Aki proposed a single-scattering hypothesis to explain the earthquake coda. According to this model, the waves seen on a seismogram are assumed to have left the earthquake source and been reflected just once en route to a recording site (Figure 4.5C). The later the energy arrives, the farther it traveled before being scattered back. According to this assumption, the envelope,

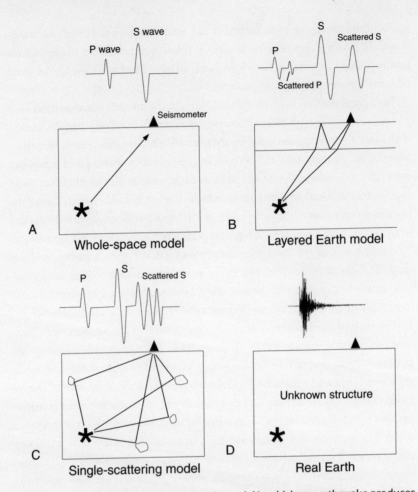

Figure 4.5. Four rupture models. *A*, a simple model in which an earthquake produces a P and an S wave; *B*, a model in which reverberations result from a layer near the surface that has a lower seismic velocity than the material below; *C*, a model in which blobs of material give rise to scattered waves; *D*, a typically complicated real seismogram, which reflects a complex combination of scattering and layering effects.

or overall gradual amplitude decay, of the coda can be related to the intrinsic attenuation within the medium. This so-called coda-Q method was instantly attractive to seismologists because it offered a method whereby a seismogram from any station could be used to estimate attenuation. Other methods, in contrast, required observations of an earthquake from a suite of stations to quantify the damping of wave amplitudes.

The coda-Q method was, to some extent, a deal with the devil. Although many seismologists were not completely comfortable with the single-scattering hypothesis, the theory was significant in providing a model for the coda and was widely applied throughout the 1970s and 1980s. Although later studies did reveal the limitations of the model—and improvements were made—the collective body of coda-Q results proved valuable. Most notably, they helped to quantify regional differences in attenuation.

In 1983, seismologists Sudarshan Singh and Robert Herrmann published a landmark paper quantifying the difference between attenuation in central and eastern North America and attenuation in the tectonically active West. That the attenuations were different came as no surprise; from almost the earliest observation of earthquake effects, seismologists had noticed that earthquakes in the East were felt over significantly larger areas than events of comparable magnitude in the West. This results from the different nature of crustal structure across North America. Wave transmission in older, colder, and less fractured intraplate regions is more efficient (has a higher Q) than in the warmer, highly faulted crust in the West. The coda-Q map made by Singh and Herrmann was the first view of the difference between eastern and western North America.

Eventually, however, seismologists moved beyond the single-scattering hypothesis. In 1987, seismologists Leif Wennerberg and Arthur Frankel proposed an "energy-flux" model in which coda envelopes were related to intrinsic and scattering attenuation. Later, seismologist Yuehua Zeng proposed a wave propagation model that was even more complex, a model that accounted for multiply scattered waves.

In addition to coda-Q studies, there were many observational studies in the 1980s and 1990s that focused on attenuation mechanisms. Seismologists asked questions like Is intrinsic or scattering attenuation more important in controlling the amplitude of strong ground motion? Do different mechanisms affect the seismic spectrum in different ways?

Attenuation might seem straightforward enough, but an enormous body of scientific literature proves that the simplicity is illusory. Nevertheless, attenuation remains a critical parameter for earthquake hazard mapping, the parameter that tells you whether shaking from a given earthquake will dampen to harmless levels within 100 kilometers or 500 kilometers. And so efforts to fully understand and quantify the attenuation of earthquake waves will likely continue for some time to come.

SITE RESPONSE

Investigations of attenuation are hampered by the last factor that significantly affects ground motions: site response. While attenuation describes the diminution of amplitudes that happens when waves travel through the crust between the fault and a site on the earth's surface, the term "site response" is reserved for the effect of wave propagation within the near-surface crust. Whereas attenuation reduces shaking, site response frequently amplifies it, sometimes in dramatic fashion. Examples of significant site response abound, from virtually every damaging earthquake in recent memory: Mexico City in 1985; Loma Prieta in 1989; Kobe, Japan, in 1995. In each instance, much of the damage was attributable to wave propagation effects in the near-surface crust.

The term "near-surface" is admittedly quite vague. Just how close to the surface is "near"? This innocuous question remains the subject of vigorous research and debate because the classification of wave propagation effects is rather arbitrary. It might be easy enough to talk about attenuation along the travel path of a wave and site response from the near-surface structure, but in fact the two effects are sometimes difficult to separate. In large and deep basins and valleys such as those where most Californians live, the distinction between path effects and site effects blurs (Figure 4.6).

The term "site response" has been used to describe the effects of shallow near-surface sedimentary layers, a few tens to a few hundreds of meters deep. Layers of shallow, unconsolidated sediments are common around the globe; they are the result of erosional processes that constantly wash minute bits of rock into river valleys, lake beds, flood plains, and so on. Within lakes and oceans, biological materials—decaying marine organisms and their ilk—can contribute significant sedimentation.

Young sediments nearly always overlie more consolidated rock. Even when sediments are deposited onto other sediments, the older layers are generally harder by virtue of the consolidation that comes with age and the increasing weight of the younger material. To a seismologist, "hard" and "soft" are quantified by a parameter known as *impedance,* the multiplicative product of seismic wave velocity and the density of a rock. Because the density of rocks and sediments doesn't vary much compared to the velocity of waves traveling through those rocks, wave velocity usually serves as a proxy for impedance in the classification of hardness.

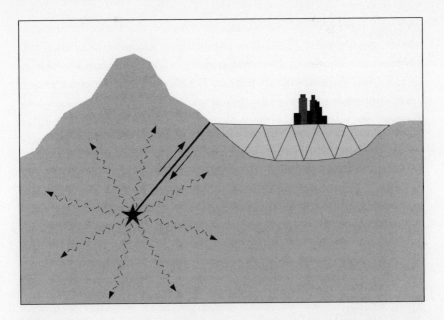

Figure 4.6. The behavior of seismic waves in a large-scale basin adjacent to mountains. After seismic waves leave a fault, they are modified in complex ways when they encounter such a large-scale basin or valley. In general, the waves become trapped and the shaking is amplified, although the actual behavior of waves is invariably extremely complicated, involving focusing, defocusing, and variability associated with the detailed structure of the basin.

As early as 1898, seismologist John Milne observed that ground motions were higher at sediment sites than on harder rock. This observation had been documented even earlier by discerning observers after the 1811–1812 New Madrid earthquakes. No fewer than a half dozen contemporary accounts tell of stronger shaking in valleys or along riverbanks. A resident named Daniel Drake wrote of strong shaking in the town of Cincinnati, Ohio, but went on to observe that "the earthquake was stronger in the valley of the river than on the adjoining high lands. Many families living on the elevated ridges of Kentucky, not more than 20 miles from the river, slept through the event, which cannot be said of perhaps any family in town."[15]

The amplification of waves within sedimentary layers can be predicted from physical principles. When seismic waves encounter a low-velocity, near-surface sedimentary layer, three things occur: the amplitude of the wave increases, the wave path is bent toward vertical, and the wave becomes trapped in the near-surface layer (Figure 4.7). All three phenomena result from the physical laws governing the behavior of waves in layered media. The first behavior can be understood with respect to energy balance. The energy associated with a wave of a given amplitude is lower in a low-velocity medium than in a high-velocity one. But wave energy is conserved when waves move from one medium to another, and so the amplitude of a wave will increase when the medium velocity decreases. In other words, when a wave slows down, its amplitude must grow.

Site response is thus associated with a singularly unfortunate double whammy: wave amplitudes are amplified and then trapped. Within a well-defined layer of sediments, not all frequencies are amplified uniformly; instead certain frequencies are amplified preferentially depending on the layer's thickness and on the seismic velocity of the waves. This phenomenon, known as a resonance effect, reflects the fact that a layer will trap energy of certain wavelengths. Just as a violin string vibrates at a certain frequency, so, too, is site response associated with a harmonic frequency. Also, as is the case with violin string vibrations, sediment-induced amplification involves higher overtone frequencies in addition to the harmonic mode. The implications of wave propa-

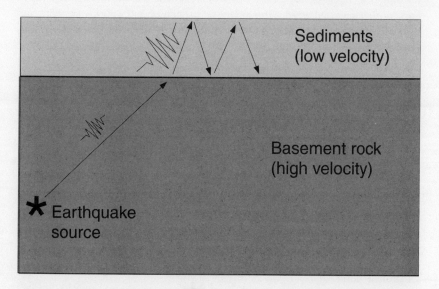

Figure 4.7. Wave amplification in a sedimentary layer. When a wave encounters such a layer, it is bent upward toward a more vertical path, its amplitude increases, and it then becomes trapped within the layer, which gives rise to prolonged reverberations.

gation in near-surface sedimentary layers are presented in Chapter 6, in the context of seismic hazard assessment.

The simple depiction in Figure 4.7 represents an idealization. In some respects, the drawing is not a bad facsimile of the truth. A sedimentary layer will behave much like the idealized model when the layer is flat and the hardness contrast with the underlying rock is sharp (Sidebar 4.2). Where the sedimentary layers are flat but the contrast is not sharp, sediment-induced amplification still occurs, but without the pronounced resonance effect. The more gradual the increase of velocity with depth, the less pronounced the resonance effect.

Sidebar 4.2   Site Response, East-Coast Style

By the late 1980s, seismologists had realized that potentially significant site response could pose a significant hazard to many areas in *eastern* North America. In many regions in the Northeast, glacial valleys filled with young, soft sediments are carved into old, and very hard, crustal rocks. The result is a more dramatic juxtaposition of geologic materials than is generally found in the West, where even "hard rock" near the surface tends to have been modified by active tectonic processes. An awareness of these issues inspired Klaus Jacob to launch one of the first tests of the ambient noise method in the United States, in Flushing Meadows, New York (see Figure 4.8a,b). The results, shown in Figure 4.8c, successfully identified a strong resonance associated with the sedimentary layers that make up Flushing Meadows. Although this particular sedimentary valley is not densely urbanized, it is home to one well-known structure that, by virtue of its size, is vulnerable to the valley resonance: Shea Stadium. If the next "World Series earthquake" occurs in New York instead of San Francisco, it could be a doozy.

When the geometry of near-surface geologic structure is intricate, simple one-dimensional wave propagation theories begin to fail. Two- and three-dimensional valleys and basins still give rise to amplifications at frequencies associated with the sediment layers' depth. Yet wave propagation becomes vastly more difficult to model. Through the mid-1980s, efforts to model site response were generally limited to one-dimensional theories. To carry the analysis to the next level of sophistication, we needed methods, theories, computer resources, and data that were previously unavailable.

Once again, the parallel explosion of disparate technologies and intellectual breakthroughs paved the way for a qualitative leap in scientific understanding. By the mid-1980s, seismologists had better data and instrumentation and began to develop the computational tools to go with them. The stage was set to bring the poetry back to ground motions investigations.

THREE-DIMENSIONAL WAVE PROPAGATION

When the M6.7 Sylmar earthquake resoundingly shook the San Fernando Valley in Southern California in 1971, it was the first earthquake stronger than M6 in the greater Los Angeles region in nearly 40 years. The prolonged lull in seismic activity coincided with a period of booming population. The Sylmar earthquake therefore captured public attention in a big way. That Southern California was at risk from earthquake shaking was no surprise at the time, but the Sylmar event highlighted the fact that hazard stemmed not only from the San Andreas Fault on the other side of the San Gabriel Mountains to the north but also from faults right under Angelenos' feet.

The event presented the seismologists with meaty new data into which they could sink their collective teeth. Seismic instruments deployed prior to 1971 recorded a profile of ground motions from the Sylmar earthquake across the San Fernando Valley. The recorded ground motions clearly reflected complicated wave propagation effects in the deep valley; the shaking of higher and larger amplitude could not be explained merely by the event's magnitude. Modeling the observations in any detail was, however, not possible with the tools available at the time.

But the earth had thrown down a gauntlet with the Sylmar earthquake and the high-quality data it produced. Here's what complex wave propagation looks like in large-scale sedimentary structures. Let's see you model it. By the mid-1980s, computer technologies and seismological methodologies had improved to the point where an assault was possible. In 1985, seismologist John Vidale developed a computer method to calculate ground motions by using wave propagation equations. This calculation had been done before, but not on a realistic crustal structure that incorporated three-dimensional complexity. Vidale employed a "finite-difference" method in which the crust is divided into a two-dimensional grid and seismic waves are tracked from one element to the next. The term "finite difference" reflects a simplification of the equations governing wave propagation, a simplification that makes the equations more tractable in the presence of complex crustal structure.

The trick to a finite-difference method is to ensure that the results are accurate despite the computational simplifications. The more computational horsepower one has, the easier the job becomes. In 1988, John Vidale and Don Helmberger tested their computer program by using strong-motion data from the Sylmar earthquake. Although their predicted ground motions did not exactly match the observed motions, the code did successfully model many of the salient features of the data, features that could not have been explained with one-dimensional models.

The decade following the publication of the Vidale and Helmberger study brought substantial breakthroughs in the analysis and modeling of three-dimensional wave propagation. Researchers improved the finite-difference code and developed new codes that used different computational approaches. At every step the application of these codes was limited by computer resources. In the mid-1980s, a single run of a state-of-the-art wave propagation program generally required hours of processor time on the best available supercomputers. In fact, so ready were seismologists with their computationally intensive codes that they were often among the first to make use of discounted or even free time that was available when research supercomputers were new and undersubscribed.

Our knowledge of wave propagation in geologic structures grew significantly through the 1980s and 1990s; progress was fueled by improved data and technology. Well-recorded destructive earthquakes have a knack for calling attention to previously unrecognized or underappreciated wave propagation effects.

TROUBLE IN MEXICO CITY

In the early morning hours of September 19, 1985, teleseismic instruments and news agencies around the globe received information that a serious earthquake had taken place along Mexico's west coast, a subduction zone well known for its ability to produce great earthquakes. News reports focused not on the situation on the coast but rather on the tragedy that was unfolding approximately 300 kilometers away, in Mexico City. Several moderate-sized buildings (six to twenty stories) had been destroyed, most of them modern office buildings; and approximately ten thousand people lost their lives. Damage along the less urbanized coast was modest in comparison.

The inference by the seismological community was immediate: the damage in Mexico City had resulted from the amplification of seismic waves within the sedimentary layers of the lake bed zone upon which the city had been built. As scientists began to analyze the data, however, a complete understanding proved elusive. A tremendous resonance effect had clearly happened within the lake bed sediments, and this effect preferentially amplified energy at approximately 2 hertz. Yet more than a decade later, the seismologists had still not succeeded in modeling the observations in detail—particularly the exceptionally long duration of the strong shaking. Although sedimentary layers will trap energy, unconsolidated sediments are also generally associated with high attenuation. The latter effect offsets the former, but in Mexico City the usual balance went awry. How and why this happened are questions that still keep investigators busy.

TROUBLE IN THE CAUCASUS

In 1988, another earthquake dramatically illustrated the importance of sediment-induced amplification. That year, December 7 became a day that would live in infamy for the citizens of the former Soviet Union territory of Armenia, when the M6.9 Spitak earthquake struck in a sparsely populated part of the Lesser Caucasus Mountains. The earthquake devastated villages close to the epicenter, where most of the structures were of a masonry design not known for resistance to earthquake shaking. But destruction on a much larger scale occurred in the town of Gyumri (formerly Leninakan), 25 kilometers to the south of the epicenter. Residents of Gyumri, a town of some two hundred thousand people prior to the earthquake, witnessed the instantaneous demise of most of the buildings in their city. Estimates of buildings rendered uninhabitable by the temblor ranged as high as 80 percent of the total inventory, which had predominantly been constructed with highly vulnerable "tilt-up" concrete walls. The Spitak earthquake claimed a heavy human toll as well. Fatality estimates ranged as high as fifty thousand souls.

Seismologist Roger Borcherdt and his colleagues demonstrated that, once again, site response was responsible for much of the damage. Several years later—an investigation was delayed for a time because war broke out between Armenia and the neighboring province of Azerbaijan—Ned Field and his colleagues demonstrated that clues to Gyumri's deadly site response could be found in recordings of ambient noise, the term for low-level vibrations caused by natural (for example, wind and ocean tides) and cultural sources. As early as the late 1970s, several seismologists in Japan had shown that at sedimentary sites, the spectrum of ambient noise sometimes reveals the dominant periods of sediment-induced amplification of seismic waves. A decade later, development and testing of the method began in earnest.

The validity of the ambient noise method remains the subject of debate. In some regions, noise and seismic signals correspond well, but in other regions the method has been less successful. It does, however, offer the enormous advantage of providing information about site response in regions where the background rate of earthquakes is low.

In the late 1990s, for example, the method was used to measure the response of the Mississippi embayment in the Memphis–New Madrid region. The intriguing results—strong amplifications at quite low frequencies (approximately 0.2 hertz)—suggest that cities along the Mississippi River might experience significant amplification in future earthquakes. Several years earlier, the ambient noise method documented potentially strong site effects in Flushing Meadows, New York (see Figure 4.8). Flushing Meadows was chosen as a test case because it is one of the many glacially carved valleys in the northeastern United States that are now filled with soft sediments.

The practical merits of the ambient noise method are clear. The generally successful results of ambient noise studies are interesting to contemplate. They imply that large-scale sedimentary structures are alive. Even the low levels of background noise in the crust are enough to excite the response of valleys and basins, albeit at low levels.

Less than a year after the Spitak event, another M6.9 earthquake grabbed the world's headlines: the Loma Prieta earthquake of October 17, 1989. No smaller than the Armenian earthquake, Loma Prieta caused far fewer fatalities (sixty-three) and, in relation to the total building inventory, far less extensive damage. Property damage totaled $6–10 billion, however, and damage to bridges and freeways disrupted traffic patterns for many years. Once again, damage was concentrated not in the immediate vicinity of the earthquake source—a sparsely populated part of the Santa Cruz Mountains—but at some distance from the rupture, at sediment sites around the San Francisco Bay. In Oakland, the filled mud flats jiggled like Jell-O, toppling not only the double-deck Nimitz Freeway but also a section of the upper deck of the Oakland Bay Bridge. The toll from amplified shaking in San Francisco's Marina district was

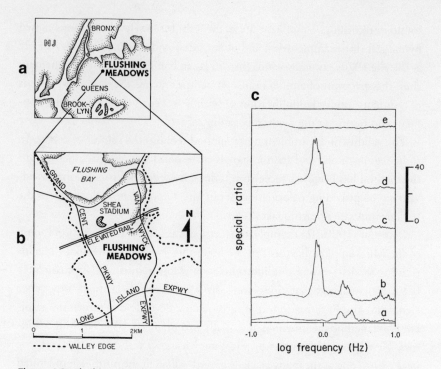

Figure 4.8. Flushing Meadows, New York, including cultural landmarks *(a, b)*; the actual site response of Flushing Meadows *(c)* estimated from recordings of background seismic noise. The sharp peaks correspond to resonance effects associated with the near-surface sediments within Flushing Meadows. Illustrations by Kazuko Nagao.

exacerbated by the vulnerable design of many quaint urban houses and apartment buildings that were constructed with living space atop a first-floor garage.

Forensic seismic investigations following the mainshock revealed that damage in Oakland and San Francisco was largely due to classic site response: a fairly common amplification in low-velocity sedimentary layers. Data from the Loma Prieta earthquake elucidated two wave propagation effects not previously understood. The first, which affected ground motions in San Francisco, is known by the ungainly name *subcritical Moho reflections*. By analyzing and modeling mainshock data, seismologist Paul Somerville concluded that ground motions in the city had been amplified by an additional factor: the reflection of waves off the boundary between the crust and the mantle. This boundary, identified decades ago and known as the Moho, is a ubiquitous feature of continental crust found at 25–50 kilometers depth. The defining feature of the Moho is a substantial rise in seismic wave velocities. The Moho's ability to reflect seismic energy upward can be predicted from wave propagation theories. These reflections had not been considered in the context of ground motions studies because, in general, waves that travel such a long path in the crust are not damaging by the time they make it back to the earth's surface. Yet Somerville concluded that subcritical Moho reflections had been generated by the Loma Prieta earthquake and had bounced energy up into the heart of the most densely populated part of the Bay Area and contributed to the damage.

The second effect illuminated by the Loma Prieta earthquake affected not the immediate Bay Area but a broader valley just south of the bay: the Santa Clara Valley, better known as Silicon Valley. Although less dramatic than the damage in Oakland and San Francisco, the damage that occurred in the Santa Clara Valley was significant and included a number of classic Spanish buildings on the Stanford University campus. This damage was, once again, largely attributed to site response caused by the wide, deep valley.

Recall that sediment-induced amplification in large, complex valleys is not generally amenable to modeling with simple theories. In the weeks following the Loma Prieta earthquake, seismologist Arthur Frankel deployed an array of four seismometers in the Santa Clara Valley to investigate three-dimensional wave propagation issues. Although deployments of portable seismometers are de rigueur after large earthquakes, Frankel's experiment was noteworthy in two respects. First, unlike most temporary deployments of portable seismometers, Frankel's used instruments that could record energy with a frequency as low as 0.2 hertz (or a 5-second period). These instruments allowed him to home in on the response of the valley, which was expected to be mostly at low frequencies (less than 0.5 hertz).

Second, Frankel grouped the instruments into an extremely tight array: stations were spaced just 200–300 meters apart. Traditional deployments of portable seismometers—always in short supply after a major earthquake—never employ spacing that tight because instruments a few hundred meters apart record largely redundant information. Indeed, the earthquakes recorded across the Santa Clara array produced similar waveforms at all stations. But that had been the goal. Because the waveforms were so similar, Frankel could track the propagation of individual wavelets across the array and, therefore, dissect the complex wave train to find out where all of the wiggles were coming from. Frankel showed that the P and S waves were arriving from the expected direction given the source location but that later waves arrived from a variety of azimuths, including some apparently scattered from the southeast corner of the valley. These late-arriving waves, known as converted basin waves (or converted surface waves), had substantial amplitude and duration. By inference, converted surface waves had played a large role in controlling Santa Clara Valley ground motions during the Loma Prieta mainshock, and will pose a major threat to the region when future large temblors strike.

Through the 1990s, wave propagation codes focused on Southern California. The preoccupation with that part of the state was largely data driven: the 1992 Landers and 1994 Northridge earthquakes each provided tantalizingly rich new data sets, including abundant recordings of strong motion within sediment-filled basins and valleys. An additional impetus to better characterize expected ground motions in the greater Los Angeles region came from the efforts of the Southern California Earthquake Center (SCEC), funded for ten years under the National Science Foundation's Science and Technology Centers Program. The SCEC comprised researchers from more than a dozen institutions in Southern California, and its mission was to support individual research efforts but also to coordinate efforts by independent groups. The legacy of SCEC is manifold, but among the center's greatest contributions was a demonstration that regionally based earthquake hazard efforts can create an effective nature laboratory. Within the "center without walls," much progress was made on regional and general issues.

Because the 1992 Landers earthquake took place in the Mojave Desert, away from the urbanized sedimentary regions of Southern California, most initial investigations of the event concentrated on matters other than wave propagation in basins. Some years later, however, the seismological community returned to the data collected across the greater Los Angeles region as valuable tools for developing methodologies to model earthquake shaking in three-dimensional sedimentary basins.

In the meantime, wave propagation issues were immediately brought to the forefront of seismologists' concerns when the M6.7 Northridge earthquake struck on January 17, 1994. Basin response had clearly been a key factor in controlling shaking, and therefore it demanded the attention of the scientific community. Scientists even joked that, by virtue of its extensive surface rupture, the Landers event had been a geologists' earthquake, whereas Northridge, which produced no surface rupture, had been a seismologists' earthquake.

The questions about ground motions that arose following Northridge were obvious. Why had the damage been so uneven across the San Fernando Valley? Why had Santa Monica, well south of the epicenter, been so hard hit? Why had one strong motion instrument near the southern edge of the San Fernando Valley, in the town of Tarzana, recorded a peak ground acceleration of twice the acceleration of gravity? Recall that an upward acceleration equal to the acceleration of gravity (1g) will throw unsecured objects into the air; shaking twice this strong is virtually unheard of.

The answer to all of these questions is that the long-period waves associated with the sizeable temblor had been influenced by extraordinarily complicated propagation effects within the three-dimensional subsurface geology. Waves leaving the source were trapped and amplified within the San Fernando Valley and Los Angeles basin, and converted basin waves produced substantial jumps in amplitude and duration. Effects varied drastically across the region as a consequence of the vagaries of wave propagation in complex structures. In some regions, unusually vigorous ground motions were caused by interference effects; other regions were hit hard by converted basin waves.

The recordings at Tarzana told a fascinating story. The instrument recording the high acceleration had been installed atop a small hill at the edge of a large valley. Instruments deployed across the hill revealed that the extreme amplification effect was localized at the top of the hill. Scientists concluded that the amplification resulted not from a shaking of the hill per se but from a focusing effect caused by the detailed geometry of the sedimentary layers below the hill.

Perhaps the most striking wave propagation story to emerge from the Northridge earthquake was not Tarzana's, which had been of little social consequence, but the story of the shaking experienced by the beach community of Santa Monica. Studies of aftershock data revealed that the amplification in this area could not be explained by simple theories of site response or even by conventional models for waves trapped in a basin. Instead, seismologist Shangxing Gao and his colleagues suggested that a highly specialized wave propagation effect had occurred. The complex geologic structure underlying the Santa Monica mountains —a dipping fault separates the mountain range from the Los Angeles basin —had acted as a lens, directing energy up into the city of Santa Monica (Figure 4.9). This focusing effect depended chiefly on the exact location of the earthquake's source. Focusing was observed in aftershocks located close to the mainshock rupture but not in aftershocks from only 10 kilometers from the source.

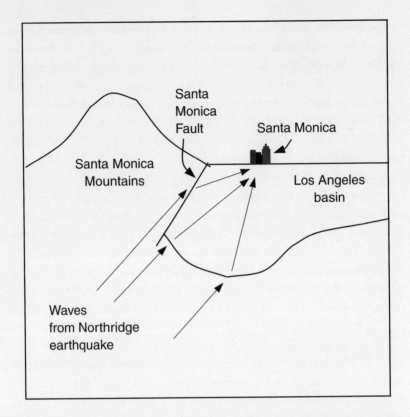

Figure 4.9. Model proposed to account for the high levels of shaking and damage in Santa Monica during the 1994 Northridge earthquake. According to this model, the geologic structure associated with the Santa Monica Fault (dipping under the mountains) produced a so-called lens effect, focusing and amplifying waves under downtown Santa Monica.

The lesson of Santa Monica is a sobering one. Surely key focusing effects will occur for other source—receiver paths throughout the greater Los Angeles area (and, indeed, elsewhere in the world). Investigating the effect once it has occurred is far, far easier than identifying regions at risk for a similar phenomenon in future earthquakes. The latter is akin to the proverbial search for a needle in a haystack. Except in this case the needle is moving around rather than sitting in one place.

Thus does a conceptual image of basin response begin to take shape. Not unlike a musical instrument, large-scale sedimentary structures will be set into motion by vibration modes whose characteristics depend on the source's location and the geometry of the basin or valley. Wave propagation effects in such environments will run the gamut from classic site response to converted basin waves to focusing effects of various types. The nature of the challenge becomes obvious. How can we hope to predict such effects for future earthquakes?

The more complicated the crustal structure, the more computationally intensive the wave propagation model has to be. By the mid-1990s, however, modeling had matured to the point that the effect of realistic structural complexities could be tested in some detail.

In the mid-1990s, seismologist Kim Olsen undertook an extensive investigation of ground motions predicted from ruptures of several major faults in and around Southern California by using three-dimensional wave propagation codes he had developed earlier. Although his codes employed a different approach than the one used by Vidale and Helmberger, his approach was conceptually similar to theirs. The Landers data had tested the method, and so it was brought to bear on ground motions from future ("scenario") earthquakes.

Olsen's results once again highlighted the importance of three-dimensional wave propagation effects. But by considering several scenario earthquakes, Olsen showed how dramatically basin response could depend not only on the crust's structure but also on the source's location and details. For example, the response of the basin to rupture of a prescribed fault segment varied on the direction that the earthquake was assumed to rupture. (The direction in which earthquake ruptures will proceed on a given fault is generally not known ahead of time.) Some generalizations could be drawn from Olsen's results. Basin response for sources well outside the Los Angeles basin was broadly similar to the response for sources immediately adjacent to the basin. Even so, variable responses within these two groups were evident.

The development and testing of wave propagation codes continues. Because the seismological community is only beginning to predict ground motions in crustal models that incorporate a realistic degree of structure and source complexity, considerable work remains to be done.

California does not have a monopoly on sedimentary basins and valleys. Although seismologists have developed and tested their methods using the abundant data from the Golden State, their results are valuable to many other regions around the United States and the world. Within the United States, researchers have documented the potential for substantial sediment-induced amplification effects in parts of the Pacific Northwest and in the Charleston,

South Carolina, area, among others. A repeat of the 1811–1812 New Madrid sequence would result in substantial amplifications in the Mississippi embayment as well as a web of amplification effects along major river valleys mid-continent. Sediment-induced amplification effects, like earthquakes in general, are by no means only a California problem.

A FINAL NOTE ON POE

Poe's sonnet reflects a not uncommon sentiment regarding science, that it is an entity whose "wings are dull realities." Art is art and science is science, and never the twain shall meet. What scientists understand—but sometimes fail to communicate—is that science has an artistry that can appeal to, not prey upon, the poet's heart.

In modern research on ground motions, artistry involves not only ideas but also visual imagery. At scientific meetings, presentations of ground motions results sometimes necessitate the use of computers to display results that cannot be easily shown on static, paper images. By the mid-1990s, such displays were commonplace in ground motions sessions in particular, as seismologists began to construct animated computer sequences to display their results. At many meetings, one can watch little movies illustrating the richly complicated dance of wave propagation in a three-dimensional crust. These models display not snapshots of ground motions as recorded by a sparse set of seismic instruments or as predicted at a discrete set of points, but the actual motions across complex geological regions, with wave amplitudes depicted by undulating ripples of color. The displays are fascinating, even captivating. Earthquake waves sometimes do amazing things, and here again the observational and theoretical advances of the late twentieth century have enormously improved our knowledge.

Edgar Allan Poe may have been an enormously gifted writer, but he was off base in his portrayal of science. Even in the seemingly pedestrian business of ground motions seismology, there is poetry to be found—poetry in motion.

FIVE THE HOLY GRAIL OF
EARTHQUAKE PREDICTION

In effect, we have redefined the task of science to be the discovery of
laws that will enable us to predict events.

—**STEVEN HAWKING,** *A Brief History of Time*

The chief difficulty Alice found at first was in managing her flamingo:
she succeeded in getting its body tucked away, comfortably enough,
with its legs hanging down, but generally, just as she had got its neck
nicely straightened out, and was going to give the hedgehog a blow
with its head, it would twist itself round and look up in her face, with
such a puzzled expression that she could not help bursting out laugh-
ing, and, when she had got its head down, and was going to begin
again, it was very provoking to find that the hedgehog had unrolled
itself, and was in the act of crawling away: besides all this, there was
generally a ridge or a furrow in the way wherever she wanted to send
the hedgehog to, and, as the doubled-up soldiers were always getting
up and walking off to other parts of the ground, Alice soon came to
the conclusion that it was a very difficult game indeed.

—**LEWIS CARROLL,** *Alice's Adventures in Wonderland*

Alice's assessment of croquet in Wonderland is almost equally true for the "game"
of earthquake prediction, an observation that appears to be at odds with Steven
Hawking's words on the very purpose of science. If the validation of a scientific
theory involves predictability and we cannot predict earthquakes, then how valid
is the science? What good are seismologists if they can't predict earthquakes?
    Such questions can be answered in part without any consideration of earth-
quake prediction per se. That is, although we cannot predict the specific time

and location of earthquakes, seismology by no means lacks predictive capability. Ground motions have now been predicted with some confidence for future ruptures of some fault segments. Such predictions mitigate earthquake risk because they enable the structural engineer to model the response of a building to ground motions it has not yet experienced. Progress has also been made in the business of long-term seismic hazard assessment, the methods with which the average long-term hazard for a region is quantified (see Chapter 6). Seismologists can also predict the distribution of earthquake magnitudes expected for a region over the long term (the *b*-value; see Chapter 3) and the rate and decay of aftershocks following a mainshock.

But what of earthquake prediction? It may not be the ultimate test of seismology's validity, but certainly it is a holy grail for the public and, yes, scientists alike. What happened to the cautious optimism of the 1970s that reliable earthquake prediction might be just around the corner? Why did the scientific community move from collective hopefulness to what can fairly be characterized as collective pessimism? Why do some members of the seismological community now argue that earthquake prediction will never be possible, regardless of any future advances in understanding?

EARTHQUAKE PREDICTION ABCs

To answer these questions, we must start at the beginning. What, exactly, do we mean by "earthquake prediction"? What are the physical requirements for prediction to be feasible?

Earthquake prediction, as the term is commonly used and understood, is the prediction of the specific time, location, and magnitude of an earthquake. Without the requirement for specificity, one could make predictions with near certainty. There will, on any given day, be several earthquakes of at least M2 in California; there will, in any given week, be several earthquakes of at least M5 somewhere on the planet. One can also predict with confidence that the San Andreas Fault will rupture in M7–8 events some time over the next 500 years.

Although such predictions are clearly too vague to be useful, the degree of specificity of proposed prediction schemes can be surprisingly difficult to evaluate at times. Certain prediction schemes have sounded fairly specific but upon close inspection have proven to be less so. The apparent success of some prediction schemes can be plausibly explained by nothing more than random chance, given the average rate of earthquakes. The true test of an earthquake prediction scheme is that it be more successful than predictions that succeed on the basis of random chance.

What are the physical requirements for a successful earthquake prediction scheme? First, earthquakes must have precursors—physical changes in the earth that occur before large earthquakes. Second, such changes must occur *only* before large earthquakes. Finally, such changes should occur before all, or at least most, large earthquakes.

Even without in-depth knowledge of earthquake rupture, we have some reason to believe that catastrophic rupture would be preceded by precursory activity of some sort. Although we don't understand how earthquakes are triggered and begin, or "nucleate," one suggestion has been that the flow of water deep in the crust might play a role. Moving fluids generate electromagnetic signals that may be observable. As an alternative, what is known as classical damage mechanics has been proposed as a model for the failure processes that culminate in earthquake rupture. Damage mechanics describes the broadly similar qualitative and quantitative processes that precede failure in any number of materials and systems—the tree that cracks and groans before it falls, the pane of glass that cracks and then shatters.

The concept that catastrophic damage is the culmination of smaller events is a familiar one. Every driver knows that an aging automobile rarely blows its engine until more minor (but costly) mechanical failures have happened. If you've ever hung on to an aging vehicle, you can appreciate just how tricky it is to predict when something will break, given the mercurial failure of the system. How much life is left in an engine or transmission that is showing signs of age? Do I invest four hundred dollars in a new clutch or two hundred dollars in a new water pump if the overall viability of the vehicle is questionable?

If automobiles are vexingly inscrutable in their wear processes, they are far simpler beasts than tectonic faults. With faults, you don't have the luxury of tinkering under the hood to see what's what. You don't have scores and scores of previous failures on which to base your analysis. Seismology is, in many ways, a nonrepeatable science—seismologists get to work only with earthquakes that decide to happen and, usually, only with the instrumental observations made at the time the events took place.

With such non-negotiable limitations, most earthquake prediction research has been directed only to the first required element for a prediction scheme: documented cases of physical anomalies that have happened before large earthquakes. Such anomalies include changes in the rate of small earthquakes and changes in the velocities of seismic waves in the vicinity of an eventual mainshock. If these physical changes occur and can be associated with the earthquake that follows, the change is said to be precursory, and it therefore may have predictive aspects.

Since there is some reason to believe that precursors exist and that we might be able to measure them, the quest to document precursors is a reasonable one. Observational Earth scientists are pretty good at measuring properties of the earth's crust; it's what we do. At any time, an area might be monitored with seismometers, GPS instruments to measure slow ground deformation, and assorted instruments to measure atmospheric conditions, stream flow, magnetic field strength, and so on.

Because the scientific community is doubtful that earthquake prediction is on the horizon, finding financial support for monitoring efforts that are focused specifically on predictions is difficult. Every so often, though, a scientist monitoring a physical quantity will be struck by an anomalous signal that occurs immediately before a significant earthquake. Such after-the-fact identification of precursors goes on outside of the scientific community as well. After any large earthquake, people will always come forward with tales of how their cat Fluffy was acting bizarrely right before the earthquake struck. Fluffy clearly sensed some signal that let her know that a temblor was on its way.

The problem, as anyone who has owned a cat knows well, is that cats act in bizarre ways for reasons unknown to humans. If the 30 million house cats in the United States behaved strangely just one day each year (a very conservative estimate!), then some seventy-five thousand cats in the country will be acting bizarrely on any given day—to say nothing of the country's dogs.

Scientific training teaches one to view one's cats with a healthy measure of skepticism. If an anomalous signal is noted, a scientist would—or should—take great pains to prove that the anomaly is truly unusual. A fluctuation in magnetic field strength that is twenty times the largest ever recorded in 50 years, for example, would be more significant than a fluctuation that has been observed once a year.

Let's suppose that a scientist can document that a precursor is genuinely exceptional. Such an observation remains circumstantial at best unless it can be plausibly linked to earthquake rupture via a sound theory. But given the uncertainty regarding earthquake ruptures, finding a plausible theory to explain any physical anomaly is not so difficult. Invariably one is left with an intriguing observation but little definite proof that an anomalous signal was indeed a precursor. Examples of such instances include highly unusual electromagnetic signals coming from the ground before the 1989 Loma Prieta earthquake documented by physicist Anthony Fraser-Smith and changes in the chemical composition of groundwater before the 1995 Kobe, Japan, earthquake. Yet recall the necessary conditions for a viable earthquake prediction scheme: a precursor must exist not only before one earthquake but before all— or at least many—earthquakes. As individual observations, such cases perhaps illuminate the nature of ruptures, but they are of limited value for prediction.

Many other precursors have been proposed over the years, only to pick themselves up and crawl away on closer inspection, like Alice's hedgehogs. Some of the optimism about prediction in the 1970s hinged on observations that earthquake wave velocities in the earth changed in a certain way before major earthquakes. In particular, the ratio between the P-wave velocity and the S-wave velocity (the V_p/V_s ratio) changed. This observation appeared to be consistent with theoretical predictions of how crustal rocks should behave under near-critical stress levels. But the measurement of V_p/V_s, like so many other quantities in seismology, is fraught with peril. For example, since V_p/V_s is a quantity that can vary spatially in the earth's crust, the ratio might appear to change over time when in fact the relevant change is the locations of the earthquakes used in the analysis.

After decades of hunting, seismologists have identified only a single common earthquake precursor: foreshocks. A certain percentage of earthquakes— in California, almost half—are preceded by one or more smaller events that occur within 3 days in the immediate vicinity of the mainshock hypocenter. Foreshocks don't occur before all earthquakes, but they occur commonly enough that every small event provides a certain heads-up that a larger event might be coming. The problem with foreshocks is that although they do occur before a substantial percentage of earthquakes in California, a much smaller percentage (about 6 percent) of small events will be just foreshocks. The occurrence of a small event therefore only slightly raises the odds of a subsequent event: 94 percent of all small earthquakes (nearly nineteen out of twenty) will not be followed by a larger event. It would be different if foreshocks somehow distinguished themselves from other events, but, with one possible exception to be examined later, foreshocks appear to be nothing more than garden-variety earthquakes. Their source characteristics and ground motions are indistinguishable from those of any other earthquake.

IS PREDICTION IMPOSSIBLE?

The indistinguishability of foreshocks from other small events may provide insight to the physical processes associated with earthquake rupture. This possibility has led some Earth scientists to the nihilistic view that earthquake prediction will never be possible. Proponents of this school of thought hold that the earth's entire crust is in such a delicate state of balance that the timing and size of events are matters of chance. Whether a small event will result in a larger event is also only a matter of chance. This model is based on what is known as the theory of self-organized criticality (SOC), which was developed from broader scientific theories that took shape in the 1970s and 1980s. The theories were based on a new mathematical concept known as chaos theory, which describes the behavior of systems that appear highly disordered but can be described by mathematical equations. (Not all disorder is chaos; sometimes a mess is just a mess.)

Theories of self-organized criticality evolved to describe multicomponent systems whose behavior is controlled by the interactions between a large number of elements, systems such as sand piles, forest fires, weather, and, some have claimed, earthquakes. The "self" in SOC reflects the fact that many systems will naturally and inevitably evolve toward a critical state—a state in which even small perturbations will disrupt the system. The hallmark of these chaotic systems is the unpredictability of their short-term behavior, although characteristics of their long-term behavior can be quantified. If you disturb a grain of sand in a sand pile, the result could be either a tiny or a massive cascade of other grains. Because the system's elements are too numerous to account for, there is no way to predict how big the landslide will be (Figure 5.1). Chaotic behavior is not restricted to systems with countless small particles; even systems with a small handful of elements can exhibit chaotic behavior because of the infinite possibilities for interaction between the elements.

Advocates of earthquake nonpredictability point to evidence that the earth's crust is indeed in a critical state and that the sand pile is therefore a good model for earthquakes. Although not granular, the crust and its faults comprise a set of innumerous elements whose interactions will control the unpredictable behavior of the system. Scientists who hold this view often point first to the distribution of earthquakes in a region. Over the long term, the *b*-value distribution—one with progressively more smaller events than large ones—will characterize the distribution of earthquake magnitudes and landslides in

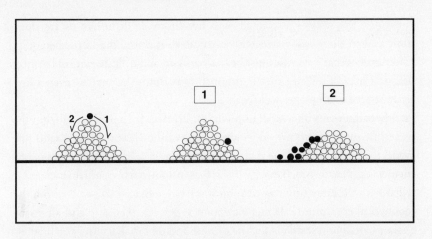

Figure 5.1. The behavior of sand pile landslides. *Left,* a grain is poised to fall in one of two directions. Although one might suppose that the results would be similar no matter in which direction the grain fell, the outcome in fact depends on extremely fine details of the sand pile shape. For example, if the grain landed in a small niche *(center),* it might trigger no further disruption. But if there was no niche for it to fall into, a much larger cascade might result *(right).*

a sand pile (and events in many other chaotic systems). That is, both distributions will manifest vastly more small events (earthquakes or landslides) than large ones, a relation which can be described by a simple mathematical equation. It was precisely the (allegedly) universal observation of earthquake *b*-value distributions that first led mathematicians and then geophysicists to think of earthquakes as the manifestation of a chaotic system.

Since such concepts were first introduced in the 1980s, other evidence has been proposed, including the phenomenon known as *induced seismicity,* whereby earthquakes are triggered by human-made disturbances of the crust: reservoir loading or unloading, oil extraction, injection wells, large-scale mining operations. Induced earthquakes are usually small, but the fact that human activity can cause earthquakes at all is a matter of concern. Proving that larger (M5+) events were induced can be difficult, but some have argued that earthquakes as large as M6 have been triggered by human activity. Some even maintain that induced earthquakes may contribute an appreciable percentage of long-term earthquake hazard in regions such as eastern North America, where naturally occurring events are rare.

Because the stresses associated with human-made disturbances are small compared to those associated with earthquake rupture, the unpredictability school argues that the crust must be in a constant and delicate state of balance, that is, a state of self-organized criticality. Disturb the balance by even a tiny amount, and earthquakes will result.

Other arguments in favor of unpredictability have been put forward, but the two I just summarized provide a good sense of the viewpoint. The sand pile model for earthquakes is one that appeals most to geophysicists who are mathematicians at heart (Sidebar 5.1). It was first introduced by mathematicians familiar with the emerging theories and was later embraced by some within the geophysical community. If this viewpoint is correct, then any and all earthquake prediction research is a waste of time and money. Earthquakes, by their very nature, will never be predictable.

One should be careful to not paint the earthquake prediction debate in overly stark tones of black and white. Although scientists are known to argue their points of view with passion (and therefore sometimes make the issue sound quite black or white), the reality in science almost always comes in subtle shades of gray. In this matter, the shades of gray constitute a softening of the absolutes. That is, few geophysicists argue that prediction has been proven impossible; instead, they believe that earthquake prediction research is not likely to be fruitful and should be given much lower priority than research on ground motions and long-term seismic hazards. On the other side of the coin, while some scientists are skeptical about arguments for unpredictability, few, if any, would disagree that the earth is a complex system whose behavior undoubtedly is chaotic and that prediction is therefore likely to be exceedingly difficult. In the end, this camp also argues that research should be directed toward questions on which progress is more certain.

The consistency of the bottom line does not belie the magnitude of the debate. In their hearts, some geophysicists are convinced that earthquake prediction will never be possible. Others, if not hopeful per se, at least refuse to subscribe to the theory of outright impossibility. As we strive to understand how earthquakes are triggered and how ruptures grow, perhaps we will come to understand that the processes are associated with precursory changes that can be measured. Or perhaps, as is so often the case in science, revolutionary change will come from a serendipitous observation that scientists today cannot even imagine. With the field of seismology scarcely a century old, imagining that the earth has exhausted its bag of tricks defies common sense.

Sidebar 5.1 Geophysicists as a Mixed Breed

Geophysicists are mutts. Unlike physicists or chemists, who usually remain immersed in one discipline for their entire careers, geophysicists are almost by definition generalists. Few universities offer undergraduate degrees in geophysics—much less in the specific field of seismology—because, as a former professor of mine is fond of saying, the earth is not a scientific field. Students therefore generally arrive at a graduate geophysics program with degrees in one of the classic fields of science (typically physics or geology) or in mathematics. Graduate school broadens one's academic horizons, usually with more coursework in mathematics, time series analysis, statistics, and the physics of the earth (for example, geomagnetism and seismology). Collectively we might be mixed-breed creatures, but most geophysicists retain some allegiance to—or at least some reflection of—their academic heritage. Such diversity is an enormous collective asset because markedly different perspectives are brought to bear on any given geophysical problem. The diversity also provides a system of checks and balances. For example, those who are mathematicians at heart will introduce and argue for rigor, while those who are geologists at heart will drag their colleagues back down to earth by insisting that models not be so mathematically elegant that they neglect real complexity. The whole is, delightfully, greater than the sum of its parts.

IS PREDICTION POSSIBLE?

One of the arguments put forward by those who are not yet willing to proclaim earthquakes to be intrinsically unpredictable is a simple one involving our limited understanding of earthquake nucleation and triggering. If we do not yet have solid, established theories describing how an earthquake is triggered and then grows, how can we say that such processes will not be accompanied by observable precursors? The timing and location of earthquakes may be unpredictable, but perhaps earthquake rupture involves more than an abrupt break that comes without warning. Perhaps a subtle, slow nucleation process plays out over hours, days, or even years before a large earthquake rears its ugly head. Perhaps if we looked closely enough at the right things, we might catch earthquake nucleation in the act.

Advocates of this argument point to a body of supporting evidence, both observational and theoretical, that they themselves have collected. Theoretical evidence comes from computer modeling of faults and crustal behavior. Modern computer models can now incorporate the complex laws of friction for the crust, laws that are similar to those derived from laboratory experiments on rocks (Sidebar 5.2), as well as simulate the crust's elasticity and the subcrustal lithosphere's viscoelasticity. Such models also allow a seismologist to circumvent the vexing nonrepeatability of real Earth processes. Unlike real earthquakes, computer simulations can be run over and over under an array of conditions.

Sidebar 5.2   Friction

Friction is another notorious geophysical hedgehog. The concept that frictional resistance between two sides of a fault must be overcome for an earthquake rupture to take place appears to be conceptually straightforward, but the nature of the resistance has proven to be far more complex than first imagined. A single coefficient of friction between two materials cannot be defined. The resistance is a dynamic quantity that is different when the fault is moving than when it is at rest, a quantity that changes in complicated ways as a fault starts to slip. Many of the materials that make up the earth turn out to be what scientists call velocity weakening. That is, the faster a fault slips, the lower the friction becomes. There is also evidence that, under some circumstances, Earth materials are velocity strengthening—that is, friction increases as a fault starts to slip. A significant challenge associated with modeling earthquake rupture is the incorporation of appropriate friction laws.

Several modeling studies in the 1990s revealed that gradual nucleation precedes an eventual earthquake rupture. On a fault that is broken into many small elements, the slip rate of a few elements in the vicinity of the hypocenter will typically start to increase months or years before the mainshock hits. We may think of a fault as locked—not slipping at all between earthquakes—

but in these simulations, elements of a locked fault can slip at low rates. When stress starts to build to a critical level and the slip rate rises, small earthquakes—foreshocks—occur in the models.

Though clearly intriguing, the interpretation of theoretical models is inevitably fraught with uncertainty. A computer program, no matter how sophisticated, is in many ways no match for the richly layered complexity of the real earth. The validity of the friction laws obtained from the models, although patterned after laboratory results, has not yet been definitely established, much less the validity of extrapolating the behavior of rock on a small scale to the large-scale processes in the earth. The division of the crust into small, discrete elements (a necessary step for computer modeling) is also a thorny issue. The behavior of models can depend on the size of the considered elements relative to general fault dimensions. Is the earth best modeled as a mosaic of large elements corresponding to different rock units or to a geometrical complexity of faults, or both? Or is it best modeled as a continuum with more smoothly varying properties? How do you model the pulverized rock (fault gouge) that makes up the fault zone itself? How do you account for the effect of water in the crust?

In light of such fundamental uncertainties, Earth scientists generally look to computer modeling not for definitive answers but for insights that might guide future observational efforts. If, for example, a model predicts an amplitude and duration of pre-event slip on a fault patch, scientists might think carefully about whether and how such a precursor could be observed.

Although models based on different methods exhibit different behavior, one consistent result emerges: precursory slips associated with nucleation are small in amplitude. Direct observation of the deformation (strain) associated with such slips will therefore be difficult. Imagine a small patch of fault less than 500 meters square and at least 10 kilometers deep slipping at a rate of 1 centimeter per second. Such movement would result in a strain signal in the surrounding crust, but the amplitude of this signal would be difficult for even the most powerful monitoring instruments to detect.

Especially if one doesn't know ahead of time what part of the fault to watch, the hunt for precursory strain signals is a lot like the search for the needle in the proverbial haystack. In one exceptional spot along one exceptional fault, however, seismologists believe their odds are better: the tiny town of Parkfield on the San Andreas Fault in California. In the 1980s, seismologists Bill Bakun and Allan Lindh concluded that six similar, quasi-periodic M5.5–6 earthquakes had occurred on the San Andreas fault at Parkfield since 1857

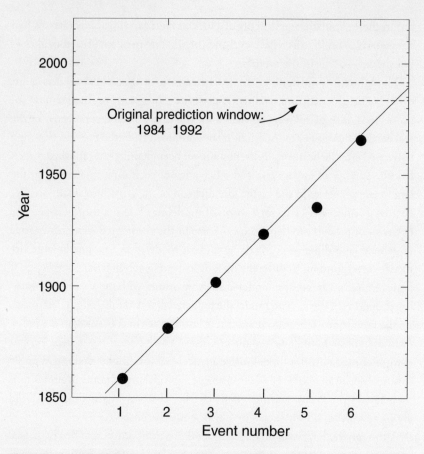

Figure 5.2. Dates of M5.5~6 earthquakes at Parkfield, California, and the original prediction window: 1984–1992.

(Figure 5.2), leading scientists to view the location as a natural earthquake prediction laboratory.

Before delving into the Parkfield experiment, let me return to the general issue of observational evidence for earthquake precursors to conclude the examination of several issues related to prediction at Parkfield and elsewhere. Precursory strain itself is difficult to catch, but the consequences of precursory strain may be observable. The primary consequence would likely be foreshocks triggered by the growing rate of slip on a fault patch. And while individually these foreshocks would exhibit no defining characteristics, the distribution of events in space and time could prove diagnostic.

Recall the basic tenets of earthquake interactions: any earthquake disturbs the crust by creating short-term dynamic stresses associated with seismic waves and long-term static stresses caused by the movement of large parcels of subterranean real estate. When events happen close in space and time—either mainshocks and their aftershocks or related mainshocks—the stress change from the first event usually triggers or encourages subsequent events. But what if adjacent earthquakes are triggered not by each other but by an underlying common physical process? In such a case, a series of foreshocks might not be sequentially triggered by the preceding event. Instead, the stress change caused by any one event might push the adjacent patch away from rupture, but that patch would rupture anyway as the underlying nucleation process continued.

This conclusion was reached for one notable recent foreshock sequence: twenty-two small events that occurred for roughly 8 hours before the 1992 M7.2 Landers earthquake. Greg Beroza and Doug Dodge concluded that this interpretation was most consistent with the results of their investigation, although the level of uncertainty associated with the results precluded proof beyond a shadow of a doubt.

The Landers foreshock sequence was noteworthy in another respect, namely, the tightness of its clustering in space and time. Recall that the Landers earthquake took place approximately 2 months after the M6.1 Joshua Tree earthquake, which was followed by an extensive aftershock sequence of its own. The twenty-two recorded foreshocks to the Landers earthquake were more tightly clustered spatially than any other cluster of events within the Joshua Tree aftershock sequence.

Another notably intense and clustered foreshock sequence led to one of the few successful earthquake predictions ever. Although insufficient data exist for this sequence to be analyzed in the same detail as the Landers foreshock sequence, a dramatic and escalating sequence of foreshocks led officials in Manchuria, China, to issue a warning on February 4, 1975, that a large earthquake was imminent. Villages were evacuated and a M7.3 earthquake struck at 7:36 P.M. that same evening. The prediction was undoubtedly not a fluke, and it saved lives. Proponents of such phenomena as animal sensitivity to impending earthquakes often point to the 1975 Haicheng earthquake; anomalous animal behavior, such as snakes coming out of the ground, was one of the observations that convinced officials to issue their prediction. Yet the Haicheng earthquake was preceded by extensive foreshocks, and many have argued that the repeated tremors alone could easily have agitated animals. Whatever one can and cannot argue about the Haicheng prediction, one critical point is indisputable: when a M7.6 earthquake occurred in Tangshan, China, 14 months after the Haicheng earthquake, no prediction had been made, and upwards of a half million people lost their lives. Whatever precursory activity happened at Haicheng was clearly not a universal phenomenon.

If prediction methodology based on immediate foreshocks is developed, it will likely be useless for the half (or so) of all events that are not preceded by smaller events. And the jury remains out on the question of nucleation. But the clustering and the stress-triggering results suggest that although each fore-shock might be nothing more than a garden-variety earthquake, the spatial and temporal distribution might be diagnostic and hence potentially useful for earthquake prediction.

Such observations do not generate wild enthusiasm within an Earth science community that comes by its humility from years of sometimes humbling experiences. The observations do provide a ray of hope—to some if not all Earth scientists—within the general climate of discouragement.

Another reason for hopefulness is a theory that emerged in the 1990s from an unexpected direction: the very same SOC theories that have led some to argue that prediction is inherently impossible. This new theory, called the critical point model and first advanced by Charlie Sammis and his colleagues, proposes that, apart from any nucleation that may play out in the immediate vicinity of a hypocenter, a large earthquake will be preceded by a preparation process that involves the full extent of the eventual fault rupture. This proposal is based on a model that considers the interactions of a multielement system; however, the model includes a twist on the SOC concept. The twist is that, unlike a sand pile subjected to the constant driving force of gravity, Sammis's critical point model includes a time-dependent driving force that controls the system's behavior. The inclusion of this force was the result of an objection that had been raised almost as soon as SOC models were proposed: the crust is not a sand pile. The slow, inexorable process of plate tectonics provides a driving stress that will, in the absence of earthquakes or aseismic release, build over time.

Recall that according to H. F. Reid's elastic rebound theory, stress builds along a locked fault as plates move; earthquakes occur when the stress overcomes the resistance (frictional or geometrical), and then the cycle begins anew. Reid might not have imagined how complicated, how chaotic, the earth's crust could be, but we need to be careful not to throw the baby out with the bathwater. Although we know that fault segments do not generate earthquakes like clockwork, we also know that the average rate of earthquakes on a fault correlates well with the long-term rate of stress, or slip, applied to that fault. The San Andreas Fault, which slips at approximately 35 millimeters per year, produces earthquakes along its major segments every few hundred years. The Mojave Desert faults that produced the 1992 Landers earthquake, faults that slip at approximately 10 millimeters per year, produce large earthquakes every few thousand years. The Seattle Fault, which slips at perhaps half a millimeter per year, might produce a large earthquake every 5,000 years.

Sammis's critical point model incorporates slip rate explicitly, and fascinating behavior results. The model reveals that the earth is not in a constant state of self-organized criticality. Instead, large earthquakes relieve the stress and move the crust away from a critical state. As stress starts to accumulate, a critical state starts to be re-established. Eventually, the stress increases and, in a sense, coalesces to the point that a large rupture can be sustained. The crust's approach to a critical state is heralded by a higher rate of moderate earthquakes in the vicinity of the eventual mainshock rupture (Figure 5.3, *top*). Scientists describe this situation mathematically by saying that the moment-release rate increases exponentially as the crust approaches a critical state.

Exponentially increasing moment-release rates had been observed before the development of the critical point model. For example, Charles Bufe and David Varnes analyzed the rate of moderate earthquakes in the San Francisco Bay Area in the 1970s and 1980s and discovered a sizeable rise in the event rate during that time, a rise that culminated in the 1989 M6.9 Loma Prieta earthquake. Since that time, other researchers have substantiated Bufe and Varnes's conclusion with data from other large earthquakes that struck the Bay Area and other regions.

The critical point model has made several key contributions to the field of earthquake prediction. First, the model provides a theoretical framework within which the observational results can be further explored and better understood. Second, Sammis and his colleagues have bolstered the evidence for exponentially increasing moment-release by demonstrating that, as the model predicts, the rate increase occurs quite generally, over a region that reflects the size of the eventual mainshock rupture. Third, the model predicts small-scale details of the process that could be central to earthquake prediction. According to the critical point model, the rate of moderate earthquakes rises not in a smooth exponential curve but in an exponential curve with smaller-scale fluc-

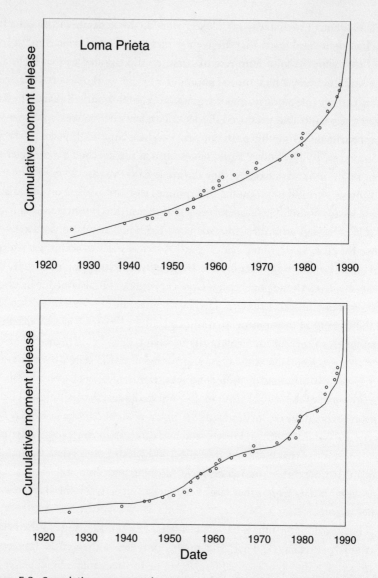

Figure 5.3. Cumulative moment release over a circular region several hundred kilometers in radius, centered at the location of the 1989 M6.9 Loma Prieta earthquake. Each circle represents an earthquake that contributed to the overall moment release. The smooth curve *(top)* shows how release continued to build prior to the Loma Prieta earthquake. The bumpy curve *(bottom)* illustrates a more complex pattern that has been suggested based on theoretical considerations. Adapted from work by David Bowman and his colleagues.[16]

tuations superimposed (Figure 5.3, *bottom*). Although it remains unclear whether these small-scale fluctuations always occur, they could, in theory, help pin down the predicted time of system failure (that is, the time of the main-shock). Although earlier approaches used the observation of exponentially increasing moment-release to predict the time to failure, these approaches were not based on theoretical models and were, as a rule, imprecise.

The jury is still out on the validity and utility of the critical point model for earthquakes. The generality of the phenomenon of increasing moment-release remains to be seen, as does the utility of such observations for short-term prediction. As of 2001, no specific predictions have been based on the critical point model; all identifications of precursory signals have been made in retrospect, after the occurrence of the mainshock. Predicting earthquakes is always easier after the fact (Sidebar 5.3). With the critical point model, the challenge is to identify a priori the region over which the accelerating moment-release is occurring. It might seem straightforward enough to simply watch for the development of higher earthquake rates in a region. But the rate of earthquakes is so low, even in tectonically active regions, that the act of identifying a statistically significant rate increase can be difficult.

Let's optimistically assume that the validity of the critical point model will be substantiated by further studies and that schemes will be developed to identify the regions and faults that are building up to large events. Even if our assumptions prove to be valid, precise temporal prediction (that is, within days or weeks) will likely remain an elusive goal. Imagining the most wildly optimistic scenario, we might look toward the day when a fault segment could be identified as ripe for a major earthquake within 1 to 5 years.

I cannot stress enough that there is no guarantee that this scenario will ever come to pass. We have an intriguing new theory supported by a measure of observational evidence, but the Earth science community has been at this point

---

Sidebar 5.3 The Art of Prediction

Very shortly after the M6.7 Northridge, California, earthquake struck at 4:31 A.M. on January 17, 1994, when the Southern California Earth science community was still abuzz with the frantic adrenaline rush that accompanies large and destructive local earthquakes, an eminent seismologist was asked if anyone had predicted the earthquake. "Not yet," came his deadpan reply.

before. Thus, we move forward with the utmost caution and care, perhaps with a hint of optimism but certainly also with too much humility to overreach by making any promises. Earthquake prediction is a puzzle, a deadly serious game not unlike the Wonderland game of croquet, with rules that are as non-negotiable as they seem to be inscrutable. Our hedgehogs might not have minds of their own, but it is often tempting to conclude that they do, so determinedly vexing and seemingly mischievous has been their behavior.

THE PARKFIELD PREDICTION EXPERIMENT

In 1984, an earthquake prediction in the United States earned an official seal of approval from the National Earthquake Prediction Evaluation Committee (NEPEC), a committee of experts charged with the responsibility of evaluating earthquake predictions. In a paper published in that same year, seismologists Bill Bakun and Allan Lindh presented a graph like that shown in Figure 5.2 and predicted that a M5.5–6 earthquake would strike on the San Andreas Fault in central California at Parkfield in 1988, plus or minus 4 years.

The graph was so striking, the implication so clear: moderate earthquakes occur at this spot along the San Andreas fault almost like clockwork, every 18–26 years. Unfortunately, the graph has gone down in the annals of seismology as one of the most notorious hedgehogs of our time.

With it, the Parkfield Prediction Experiment was born, apparently of impeccable logic. Here was a golden opportunity to catch an earthquake red-handed. In 1984, these heady developments were described in the journal *Science* in a "News and Views" article titled "Stalking the Next Parkfield Earthquake." Over the next few years, the Parkfield region was instrumented with all types of monitoring equipment, including seismometers to record big and small earthquakes, and sensitive strain meters to record the ground's slow deformation.

Over the ensuing years, researchers thought they saw signs that a preparation process was underway at Parkfield. In 1986, another "News and Views" article titled "Parkfield Earthquake Looks to Be on Schedule" was published in *Science*.

On December 31, 1992, the prediction window came to a close, not with a bang but with a whimper. A couple of moderate earthquakes, M4 or thereabouts, did take place at Parkfield within the prediction window, including a M4.7 event in October of 1992 that led to a public warning of a possible mainshock. The community held its breath, but the expected mainshock didn't happen. By 1992, the "1988 Parkfield earthquake" was clearly late to the ball. It remains to be seen now just how late it will be. (*Science* ran no "News and Views" articles on Parkfield between 1986 and 1993. However, a technical article by geophysicist Jim Savage was published in another journal in 1993; this article was titled "The Parkfield Prediction Fallacy.")

To some, the Parkfield Prediction Experiment is a shining example of a government-supported flop. We relied on a bogus prediction and spent millions to instrument the area and analyze the data, all to catch an earthquake that didn't happen.

Seismologists have come to understand that the 1984 Parkfield prediction was overly optimistic. For one thing, the apparent periodicity of earthquakes in one location might be nothing more than a fluke. But even if the regularity of the sequence does reflect the physics of the system, if you look closely at Figure 5.2 you see that the sequence has not quite exhibited clockwork regularity. The last three Parkfield earthquakes occurred in 1922, 1934, and 1966. As every second-grader knows, 34 minus 22 is 12, and 66 minus 34 is 32. That both intervals were substantially different from 22 was recognized. But struck by the greater periodicity of the earlier events, seismologists proposed that the 1934 event had been triggered ahead of schedule, which lead to the longer interval.

In the early 1990s, Yehuda Ben-Zion developed a computer model of the San Andreas Fault at Parkfield. His results showed that a sequence of moderate Parkfield earthquakes could be modeled as the prolonged consequence of the great Fort Tejon earthquake of 1857 that ruptured 350–400 kilometers of the San Andreas from Parkfield southward. The model also predicted that the Parkfield events would be quasi-periodic, but that the interevent times would increase after 1857. In light of these results, the 32 years between 1934 and 1966 starts to look less like a fluke and more like a pattern (especially now that the most recent interval is at 34 years and counting).

The community has come to another realization in the years since the 1984 prediction was made. Some of the earlier Parkfield events—the ones whose regularity so impressed researchers—might not have been true Parkfield events at all but events on fault systems adjacent to the San Andreas. With little if any instrumental data for the earlier events, sparse felt reports allow only crude locations to be inferred. At the time the events were first analyzed, it seemed obvious that any moderate events in the area would have been on the San Andreas. Yet our knowledge of fault systems immediately adjacent to the San Andreas has increased recently, particularly after the 1983 M6.3 Coalinga earthquake that occurred along the secondary fault system.

Should researchers have explored the possibility that some of the earlier "Parkfield" events had not occurred on the San Andreas Fault? Perhaps. But predicting earthquakes is like quarterbacking on Monday morning. Decisions made during the Parkfield Prediction Experiment have been imperfect, yes, but the legacy of the experiment is not—should not be—one of failure. By placing a natural laboratory in Parkfield, the community has made tremendous strides in understanding the nature of faults and earthquakes. Many discoveries have been made because of the quality and quantity of Parkfield data. The experiment might not have played out as we expected or hoped, but it was a worthwhile learning experience nevertheless. Much of the community's respect for the daunting challenge that earthquake prediction represents can be attributed to our experiences at Parkfield. If the hedgehogs are this vexing at Parkfield, just imagine what they must be like elsewhere.

And so, thankfully, the monitoring efforts at Parkfield continue. The behavior of a quasi-periodic system will, ultimately, require centuries of observation in a tectonically active region such as California. But in the meantime Parkfield is the spot on the San Andreas Fault that still represents the community's best hope of catching a M6 event; hence it remains our best hope of finding that needle in the haystack. We watch, we wait, we work, we learn.

VAN

If the Verotsos, Alexopoulous, and Nomicos (VAN) prediction method is less well known among the general public than the Parkfield Prediction Experiment, it is certainly no less infamous within the Earth science community. With a name incorporating the initials of the three scientists who introduced the method—Panayotis Verotsos, Kessar Alexopoulous, and Kostas Nomicos— VAN has generated a debate as extensive, polarized, and sometimes acrimonious as any that have played out within the halls of science. Opinion articles have been published in scientific publications; an entire volume of the journal *Geophysical Research Letters* was devoted to the VAN method and debate in 1996.

The VAN method identifies "seismic electric signals" that supposedly precede to large earthquakes in and around Greece, from whence the method's proponents hail. The VAN team has been observing such signals since the late 1970s and issuing predictions since the early 1980s.

That the seismic electric signals represent real fluctuations in the electrical field level of the region is not in dispute; their existence is well documented.

What is much less certain is what these signals mean. VAN proponents identify a small number of "sensitive sites" at which electrical signals are recorded for earthquake source regions. Yet even proponents of the method have not been able to formulate a plausible physical mechanism that could produce the observed signals. Although the lack of a physical model does not invalidate the observations—in fact, in the data-driven field of geophysics, solid observational results often lead to the development of theories—the VAN observations are especially problematic. First, although rocks can generate electrical signals when they are deformed, the amplitude of the seismic electric signals is much higher than any mechanism can account for. Some researchers have investigated VAN precursors and concluded that they were of industrial origin. That mechanical activity can generate substantial electric field fluctuations is not disputed.

The VAN signals defy physical intuition in other ways. Seismic electric signals are supposedly observed at different "sensitive sites" for different earthquake source regions. Yet such sensitive sites can be much farther away from the earthquake than other sites where no signal is observed.

Although it is difficult to argue that the VAN tenets are absolutely inconsistent with the laws of physics, the tenets certainly appear to be paradoxical and not remotely suggestive of any simple physical model. If the VAN proponents had a successful track record of predictions, then the community would jump to look for a physical explanation that must exist. This point sometimes gets lost in the shuffle: seismologists would like to be able to predict earthquakes. If they criticize prediction methodology, their criticisms are driven not by any innate desire to see a proposed method fail but by their recognition of the need for rigor and care in this endeavor.

Unfortunately, the track record of predictions based on VAN is as muddy as everything else associated with the method. VAN proponents claim to have successfully predicted twenty-one earthquakes in Greece during an 8-year interval between 1987 and 1995, when predictions were being issued on a regular basis. One problem, though, is that the predictions were issued on a regular basis indeed. During the 8-year period, predictions were issued for ninety-four earthquakes in sixty-seven prediction windows.

The duration of the time windows was itself vague. In the original 1981 paper, the VAN team claimed that their precursory signals preceded earthquakes by a few minutes to a few hours. In 1984, delays up to several days were described; by 1991 the delays had grown to a few weeks. Given vague time windows, VAN proponents claimed success for earthquakes that happened as long as two months after the supposed precursory signal. Looking closely at the space, time, and magnitude windows specified during the 8-year run of predictions, we find that two-thirds of the supposed successes were earthquakes that occurred outside of the stated bounds.

Therein lies the rub. If one issues a lot (sixty-seven) of predictions over a short time (8 years) and each prediction stretches for a substantial amount of time (a few days to 2 months), then one will obtain successes based on random chance alone in a seismically active region such as Greece.

We have by no means seen the last of the VAN debate. Proponents of the method, including some researchers other than the three who developed it, remain determined to demonstrate the validity of their prediction scheme. But if it is not quite dead in the minds of mainstream Earth scientists, VAN appears to be on life support and failing fast.

A FINAL NOTE: EARTHQUAKE PREDICTION AND PSEUDOSCIENCE

The public is rarely if ever privy to the meat of the debates that rage within the Earth science community on the subject of earthquake prediction. Seismic electric signals, self-organized criticality, statistical evaluation of a method's success—such topics are sometimes difficult for researchers to get their collective arms around, much less explain in terms that make sense to a nonspecialist audience. Moreover, virtually all communication from the scientific community to the general public is filtered through the media, a filtering process infamous for producing stories—or, more frequently, headlines and television news teasers—that leave scientists wincing.

Scientists should talk in their own voices more often. That fact that they don't reflects a shortage of vehicles apart from the media for communication between scientists and the public. It also reflects some truth behind the stereotype that those who do science are not always those who are best at explaining science in plain English. In addition, the research science community is permeated by a certain culture in which what counts (for the critical purposes of hiring, promotions, and tenure review) is research that leads to technical articles in scholarly journals. This culture is changing (outreach efforts now typically earn a scientist a few Brownie points) but only slowly (such efforts will not earn you tenure).

This muddled climate results in a void that serves as a beacon to non-specialists who are convinced that earthquake prediction is really not so hard, who are convinced that they themselves have made discoveries that have eluded the scientific establishment. The establishment then turns around and squashes outsiders' contributions by retaining its tight-fisted control of the journals and meetings by which scientific results are disseminated.

Conspiracy theories are delightful beasts. Support the establishment, and you are part of it. Point to a lack of substantiating evidence for the conspiracy, and that is turned into proof that the conspiracy has been successful. (Of course no aliens have been found in the New Mexico desert; the government hid them too well.) The truth—banal as truth is wont to be—is that the scientific establishment is far from hegemonic. With the dawning of the information age, research materials, data, and guidelines for manuscript preparation are now more readily available than ever. Our scientific journals and meetings are open to any individuals willing to meet established standards for presentation, proof, and review.

Interested nonspecialists will always, it appears, decline to do this, opting instead to shadow the mainstream scientific community with a culture all their own. That culture is a fertile spawning ground for earthquake predictions of all conceivable flavors. Predictions are regularly issued based on electromagnetic signals, bodily aches and other symptoms among sensitive individuals, weather, clouds, solar flares, and the voices of angels (Sidebar 5.4).

With the alternative being a universe that is resolutely—and sometimes tragically—random and unpredictable, some people will always rush to embrace the comfort of understanding, even when such understanding is grossly flawed. Such are the seeds of pseudoscience.

Sidebar 5.4   Angels We Have Heard on High

Of the various types of earthquake predictions that have come to my attention, those based on the word of angels or other unearthly apparitions fall into a category all their own. Science is science, and faith is faith; one has no standing to pass judgment on the other. I will, however, say that I have yet to hear about an angel whose track record on earthquake prediction was any better than that of mortal beings.

Generally discounted by the mainstream media and therefore relegated to the fringes, pseudoscientific earthquake predictions rarely make a big splash in the public eye. One notorious exception came in 1990, when self-professed climatologist Iben Browning predicted a large earthquake in the Memphis area. The prediction, based on (very small) tidal forces, aroused the public's concern; and government officials had to go to considerable expense to defuse the public's anxiety.

What is perhaps most unfortunate about the Browning prediction is the extent to which it discredited the Earth science community in the eyes of the public. The distinction between legitimate and illegitimate credentials might be clear to researchers in the field, but it is not always easy for outsiders to judge. Browning, who published a newsletter on weather and commodities prediction, appeared to have some expertise on the subject of tides and could make plausible arguments that linked tidal forces to earthquakes. To some, the arguments advanced against the prediction were the stuff of yet another inscrutable technical debate between experts. The debate, and the eventual failure of the prediction, in the end eroded the public's confidence in the Earth science community as a whole.

Worse yet, the public sometimes seems inclined to apply zero-tolerance standards to earthquake (and volcano) forecasting. A single failed prediction—such as Parkfield or a 1980 U.S. Geological Survey alert concerning an eruption at Mammoth Mountain—can be a serious setback to the entire community's credibility. While understandable on one level, the standards to which earthquake predictions are held can be frustrating. On any given day one can find reputable economists happy to publicly predict that the stock market will go up, go down, or stay level. Even a steady diet of such dissension does not appear to invalidate the science of economics in the mind of the general public.

One would not wish to apply an economics model to earthquake predictions. It is ludicrous to imagine news programs summarizing the earthquake activity of the day with a resident expert forecasting trends based on the blip of the day's activity.

What Earth scientists wish for is a good, solid question to consider. An improved understanding of the issues could be achieved, perhaps, if people recognized that communication is a two-way street. Getting to the end of this chapter—the writing and the reading both—is a small, but positive, step in the right direction.

SIX   MAPPING SEISMIC HAZARD

The essence of knowledge is, having it, to apply it, not having it, to confess your ignorance.

—CONFUCIUS

Having been born a millennium or so too early to have witnessed the development of probability theory, Confucius may have seen knowledge as an either-or proposition: you have it or you don't. With probabilistic seismic hazard analysis, however, you can apply what knowledge you have and at the same time confess your ignorance. Faced with a growing awareness of the intractability of earthquake prediction and a stronger imperative to apply hard-won knowledge to matters of social concern, the seismological community has turned away from short-term earthquake prediction and toward long-term seismic hazard assessment.

Seismic hazard assessment—also known as hazard mapping—involves not the prediction of any one earthquake but rather the evaluation of the average long-term hazard faced by a region as a result of all faults whose earthquakes stand to affect that region. Because quantifying earthquake hazards requires evaluation of earthquake sources, wave propagation, and the effects of local geology, the construction of hazard maps requires Earth scientists to integrate knowledge from disparate lines of inquiry. What faults stand to affect a region? When those faults rupture, how will the energy be dampened, or attenuated, as it travels? How will the geologic conditions at each site modify the shaking?

Earth scientists are careful to differentiate "hazard," which depends on the earthquake rate, from "risk," which reflects the exposure of structures, lifelines, and populations to existing hazard. Risk is generally the purview of insurance companies; the evaluation of hazard is a matter of science.

Hazard assessment is inherently a game of probabilities. Hazard is usually cast in terms of the probability, or odds, that a certain level of shaking will be

131

exceeded at a location over a specified time frame. The length of time chosen for hazard quantification depends on the application. If you are siting a nuclear waste disposal facility, you must worry about future ground motions over tens of thousands of years. More commonly, though, hazard is quantified over time frames closer to 50 years, the length of time corresponding, more or less, to an adult lifetime.

Hazard mapping usually quantifies not the shaking expected to occur in a 30- to 50-year window but rather the shaking that could (with odds high enough to worry about) occur within this time frame. Perhaps this distinction is a fine one, but we must understand it to make sense of hazard maps. Seismologists generally estimate the average shaking hazard from all potential earthquake sources over a 500-year interval and then present the results in terms of the hazard that is expected with 10 percent probability over the next 50 years. Such estimates are referred to as the 10 percent exceedance levels. Exceedence levels may seem confusing, but remember the math can always be reversed: hazard with a 10 percent probability in 50 years is essentially the same as expected hazard within 500 years.

The use of exceedance levels is largely a matter of convention, but it results from the disconnection between the time frame associated with earthquake cycles (hundreds to thousands of years) and that associated with human lifetimes. A map quantifying expected hazard (at 100 percent probability) over 500 years—the span of time over which an average hazard can be estimated with some measure of confidence—might seem less relevant than one quantifying the hazard over 50 years.

Seismic hazard mapping is by no means new. Probabilistic hazard maps for California and the United States as a whole were developed as early as the 1970s. These early maps were generally based on the simple premise that the background rate of small to moderate earthquakes in a region could be extrapolated to yield average rates of future large events.

To move to the next level of sophistication of hazard assessment, we must assemble a jigsaw puzzle of the pieces mentioned earlier: faults, wave propagation, and local geology. But assembling the pieces of this particular puzzle isn't as difficult as constructing each individual piece in the first place. Probabilistic hazard is determined after all of its individual components have been independently evaluated. In the latter part of the twentieth century, enormous strides were made in our understanding of earthquake sources, attenuation, and site response, strides that allowed us to hone our ability to quantify long-term hazard. Advancements in hazard mapping can be viewed as the culmination of progress in a diverse collection of earthquake science endeavors in the fields of geology and seismology. If advances in disciplines such as earthquake-source imaging and seismic wave propagation have been revolutionary, then so, too, must be the product that integrates these and other elements.

Revolutionary advancements in hazard mapping are far more complex than the derivation of a single theory or even an overarching paradigm, however. Endeavors like hazard mapping have come to be known as "systems sciences," sciences that deal with many complicated pieces and the interplay among them. Like any systems science, hazard mapping is therefore not simply a puzzle to be solved but a mélange of multiple puzzles, each to be solved in its own right and then finally fit together.

The Earth science community embarks on this arduous journey because it must. Without seismic hazard assessment, Earth science knowledge would never be applied to the critical business of hazard mitigation. Recall that hazard results from earthquakes, but risk is a function of buildings and population. We cannot do anything about earthquake hazard, but if we understand it well enough, we can do a lot to reduce the risk that earthquakes pose.

PUZZLE PIECE #1: EARTHQUAKE SOURCES

The element of hazard mapping that is arguably the most difficult is evaluating faults in a region and predicting the distribution of earthquakes they are expected to produce. The first step is to compile a list of known faults, which is no easy task because some faults are buried deep in the earth and others are frightfully complex. Then the enormously daunting task of determining the average long-term slip rates for each fault must be tackled. If scientists can determine how much movement has occurred on the faults in the past, then they can predict the future rate of movement. This information is painstakingly assembled through careful geologic investigations of past earthquakes and their cumulative effect on the earth's landforms (Figure 6.1). Such investigations consider the long-term motion across faults as evidenced by the disruption of rock formations and river channels and by paleoseismic exhumations of faults to document the number (and therefore the rate) of earthquakes in prehistoric time. Paleoseismology involves the geologic investigation of fault zones, in which evidence of past earthquakes is sometimes preserved in the near-surface geologic strata. At some sites, we can glean evidence for multiple prehistoric

Figure 6.1. The San Andreas Fault, running diagonally through the image. Drainage channels, evident as linear features running across the frame, are offset where they cross the fault, which indicates the ongoing right-lateral strike-slip motion. Photograph by Robert Wallace.

earthquakes, thereby extending our knowledge of earthquake history back by hundreds, and even thousands, of years.

Although still very much a work in progress, the quantification of slip rates on major faults in tectonically active regions has come a long way. Some of the most sophisticated analyses have been done in Southern California. In 1995, geologist James Dolan and his colleagues published a summary of the (then) current body of knowledge for the half dozen major fault systems in the greater Los Angeles region. As early as 1998, subsequent studies suggested revisions to the Dolan model. All current geologic models will almost certainly be revised in years to come, but the community is now somewhat confident that current knowledge is good enough to be applied. Moreover, we really have no choice but to apply current models. If we foreswore the application of geologic results until they were considered perfect, our knowledge would never be applied.

In Southern California, as well as other regions, plate tectonic considerations provide an element of "ground truth" to geologic models. For example, the long-term slip rate of the San Andreas Fault in Southern California has

The concept of long-term slip rate is a little confusing because most faults do not slip steadily. At any give time, the slip on a locked fault is most likely zero. Average, or long-term, slip rate represents the cumulative motion of a fault over many earthquake cycles. If a fault produces an earthquake with 5 meters of slip every 500 years on average, its long term slip rate is 1 meter per 100 years, or 1 centimeter per year. Active faults typically have slip rates in the range of 1 to a few centimeters per year.

been determined to be approximately 3 centimeters per year (Sidebar 6.1). Because the San Andreas makes a bend (the Big Bend) through the Palm Springs region (Figure 6.2; Sidebar 6.2), compression is introduced both north and south of the bend. Try holding your right hand in front of you, palm up and elbow out to the side. Now place the left palm on top of the right (left elbow out to the left side). Now try sliding your hands past each other while bending your wrists—your hands are forced into each other as you try to move them laterally. The crust in the area of the Big Bend does the same thing. To the south of the bend, the region of compression encompasses most of the highly urbanized regions of Southern California.

Given the San Andreas Fault slip rate and the geometry of the Big Bend, one can apply the principles of high school geometry (see, it *is* good for something!) to determine an overall compression rate across the greater Los Angeles region. The result compares nicely to the rate of 8.5 millimeters per year that Dolan and his colleagues estimated by considering geologic and deformation data. The consistency provides a measure of confidence in the geologic model. The consistency is also scientifically important because it implies that any aseismic processes (deformation processes that take place without earthquakes) are relatively minor; the lion's share of the strain associated with compression will be released in earthquakes.

The strain rate for a region can be thought of as the currency in an earthquake budget. A regular paycheck of strain is the income, and the earthquakes on the faults in the region are the expenses. Once we determine how much strain is being accumulated in a region, the next step is to figure out how the earth will allocate its strain budget. Although you can spend less than you earn, the earth has no choice but to eventually pay out its entire budget. Of all the

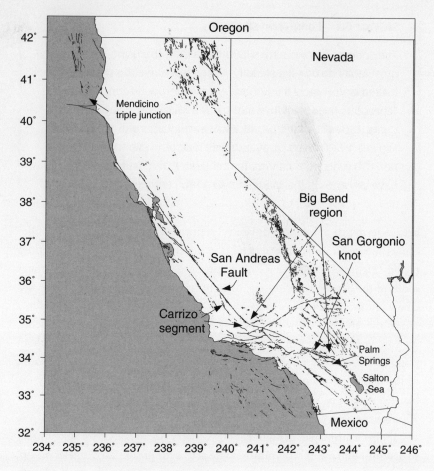

Figure 6.2. The San Andreas Fault system.

issues relevant to seismic hazard assessment, perhaps the most complex is how the earth's budget will be allocated. What distribution of earthquakes will be produced by an individual fault or a regional collection of faults?

One might imagine that answering this question would be a breeze. After all, we know what regional earthquake distributions will look like—the simple *b*-value distributions shown in Figure 3.2. Again, curiously, the financial analogy works well. Just as one has few large expenses and many small expenses, so, too, does the earth produce many more small earthquakes than large ones. Predicting earthquake distributions unfortunately turns out to be far less simple than it first appears. For one thing, the *b*-value distribution describes

Sidebar 6.2   The Big Bend

Addressing the consequences of the San Andreas geometry is relatively straightforward, but it requires that we ask a fundamental question: Why is the boundary between the Pacific and North American plates so complicated in the first place? The answer lies in the realization that, on a spherical Earth, no plate tectonic process exists in a vacuum. In the western United States, the plate-boundary system might be the biggest show in town, but it is not the only one. In particular, a large parcel of crust immediately east of northern and central California (the so-called Basin and Range area) has been undergoing a long, steady process of extension. Geophysicist Tanya Atwater and her colleagues have shown that this extension is a consequence of the subduction that took place off the coast of California before the present-day San Andreas system came into existence. Thus, as the San Andreas transports crust laterally along the coast, extension within the Basin and Range area pushes the northern two-thirds of the fault westward. Some scientists argue that over the next few million years, the San Andreas will be replaced by a more direct plate boundary that will evolve from existing faults. Any such evolution, however, will be strictly geologic in its time frame. The geologic record is quite clear that for the foreseeable future, the San Andreas will remain the biggest show in town (although not the only show).

the earthquakes expected within a region, not those expected for any fault. A *b*-value distribution clearly does not describe earthquake activity along the San Andreas Fault, for example; the fault does not produce small events at a rate commensurate with its rate of great earthquakes. Second, one cannot quantify a magnitude distribution for a region without prescribing a maximum magnitude (M_{max}), an issue that proved as contentious as any in the halls of Earth sciences departments during the 1990s. M_{max} is equivalent to a household's single largest regular expense, perhaps a mortgage payment. A family couldn't begin to understand the balance between earning and spending without knowing the dollar amount of this expense and the frequency with which it had to be paid.

The complications regarding the expected earthquake distribution for a fault largely revolve around the question of characteristic earthquakes. The characteristic earthquake model, advanced by geologists David Schwarz and Kevin Coppersmith in 1984, holds that a major fault segment will rupture in a similar manner through many earthquake cycles (Figure 6.3, *bottom*). The characteristic earthquake model was built on concepts introduced but not formalized in earlier studies by geologists Clarence Allen and Bob Wallace. The hypothesis that faults will produce earthquakes of comparable dimensions over many earthquake cycles is arguably the most conceptually intuitive of all the hypotheses that have been put forth. This hypothesis derives support from some theoretical and numerical models and from some lines of observational evidence.

Some researchers draw on other lines of evidence to suggest different models. One proposed alternative is a characteristic (or uniform) slip model, in which slip at any point along the fault is the same from one earthquake to the next, but segments of the fault rupture have varying combinations in different earthquakes (Figure 6.3, *center*). A final alternative is that successive great earthquakes are more variable still, with neither slip nor magnitude repeated from one cycle to the next (Figure 6.3, *top*). Intuitive as the characteristic earthquake hypothesis may be, it remains the subject of lively debate.

Returning to the budget analogy, we must understand the distribution of expenses, both large and moderate. Characteristic earthquakes are large events, but models to explain them also have important implications for moderate events. The San Andreas Fault does produce smaller but still damaging moderate earthquakes, such as the M6.9 Loma Prieta earthquake of 1989. Countless other faults in California can produce earthquakes that might not equal the power of the great San Andreas Fault events but are plenty damaging nonetheless. In 1994, seismologist Steven Wesnousky observed that earthquakes on young, poorly developed faults are fairly well described by the classic earthquake distribution (that is, ten times as many M4 events as M5s, ten times as many M3s as M4s, and so on). On more mature faults, though, only the smaller events are described by the simple mathematical distribution; the rate of larger events is higher than the level predicted by simple extrapolation (Figure 6.4). One can regard these larger events as characteristic earthquakes if one is willing to gloss over a minor uncertainty about the details of the characteristic earthquake and characteristic slip models. Regardless of how similar great earthquakes are, the largest earthquakes on any fault will be of com-

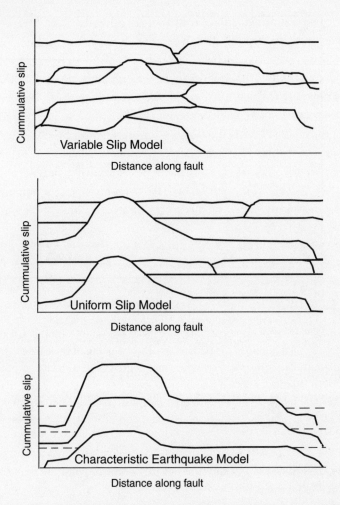

Figure 6.3. Three models for the rupture of segments of a long fault. In the characteristic model *(bottom),* identical earthquakes repeat. In the uniform slip model *(center),* any one point on the fault slips the same amount in each earthquake, but because segments rupture in different combinations, successive earthquakes are not identical. The variable slip model *(top)* corresponds to a model in which segmentation and slip vary from event to event. Adapted from work by David Schwartz and Kevin Coppersmith.[17]

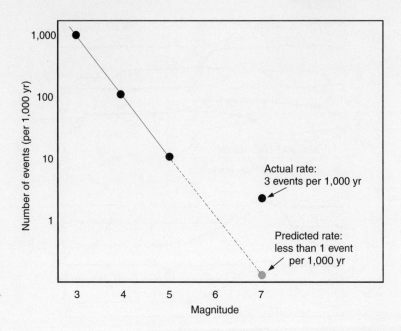

Figure 6.4. Number of events per 1,000 years on the Garlock Fault in California as a function of magnitude. On the basis of the observed rate of small events, one can extrapolate to larger magnitudes, with some degree of uncertainty, to obtain the dashed line. However, inferred rate of M7.0 events (3 per 1,000 years; *black dot*) is significantly higher than any of the predictions. Adapted from work by Steven Wesnousky.[18]

parable magnitude (M_{max}) and thus not part of the distribution that describes their smaller brethren.

These complex issues can be illustrated once again with an analogy to household budgets. A mature fault is a little like the budget of a mature household— one in which the mortgage has been paid off but which now includes occasional large expenses such as luxury vacations. If the distribution of expenses fit the *b*-value model when the mortgage was being paid, the new distribution might look quite different (Figure 6.5).

For the purposes of seismic hazard assessment, the general practice is to assume a classical (*b*-value) distribution for sources in regions away from major faults and in regions such as the eastern United States, where fault structures are poorly understood. For major faults, one generally assumes a characteristic model, not in the precise sense of the term as defined by Schwarz and

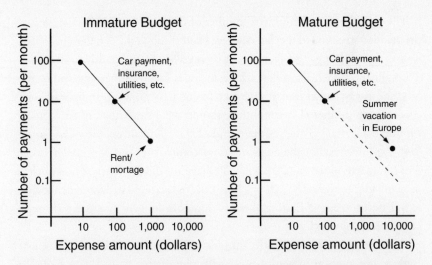

Figure 6.5. *Left,* the distribution of monthly expenses for a simplified "immature budget." An immature budget contains one large monthly expense (a rent or mortgage payment), a large number of intermediate expenses such as car payments and so on, and an even larger number of small expenses. *Right,* a simplified "mature budget." The mature budget is missing the mortgage expense but includes less frequent larger expenses. These two budgets are analogous to immature and mature faults, as discussed in the text.

Coppersmith, but in the imprecise sense of events of generally comparable magnitude.

The question of maximum magnitude remains. How does one, a priori, determined the maximum possible magnitude from a fault or region? The question is more complicated than it might sound. Given a historical record that is short compared to earthquake cycles, one cannot generally assume that the largest earthquake witnessed in historical times is the largest event possible.

Such an assumption might be more valid for California, where the earthquake cycle is short and where geologic and paleoseismic investigations have extended our knowledge of great earthquakes back a few thousand years. Because available geologic evidence—as well as direct observation of major San Andreas Fault events in 1857 and 1906—suggests magnitudes of approximately 7.9–8.0 for the great plate-boundary events, one might feel justified in assuming that this value represents the greatest possible magnitude. The truth, however, is that uncertainty still plagues our interpretations in even this case, which is far better known than most.

The uncertainty stems from a number of disparate factors. For one thing, the detailed structure of the San Andreas Fault through the San Gorgonio Pass west of Palm Springs (see Figure 6.2) is complicated and poorly understood. Through the Coachella Valley, which includes Palm Springs and Indio, the San Andreas branches into at least two major active strands. In light of such considerable geometrical complexity, should we assume that an earthquake rupture could start at the southern end of the San Andreas (near the Salton Sea), rupture through the San Gorgonio complexity, and continue another 350 kilometers north to Parkfield? We have no direct evidence that such a megaquake has ever occurred, but how can we conclude with confidence that it will never happen? Earthquakes are, after all, known to rupture through geometrical complexity.

Additionally, we must consider the possibility that the southern San Andreas Fault might rupture in conjunction with one of the major thrust faults in Southern California. Some scientists have pointed to a similar composite rupture elsewhere in the world, the M8 Gobi-Altai, Mongolia, earthquake of 1957, as a relevant analog for the San Andreas system because in both instances substantial secondary fault systems are in proximity to the main fault. Again, paleoseismic results from California suggest that catastrophic events such as these are unlikely, at the very least; the major Southern California thrust faults produce large (M7–7.5) earthquakes far less frequently than does the San Andreas. For the purpose of probabilistic hazard assessment, though, the question is not whether such an event is likely but whether it is possible.

The issue might appear to be irrelevant. If a catastrophic, "doomsday" earthquake rupture scenario were possible but improbable, then does it matter whether or not such an eventuality is considered? It does matter, a lot. Imagine knowing with certainty that an expense will come up infrequently—perhaps just once in a lifetime—but will be very large compared to other expenses (college tuition comes to mind). That single expense would have to be considered carefully in the total budget because, although infrequent, it would be both enormous and inevitable. Catastrophic earthquakes have much the same effect.

To understand this issue in detail, we have to revisit concepts introduced earlier: first, the concept of an earthquake budget and, second, the logarithmic nature of the magnitude scale. If we assume a maximum magnitude of 8.1 for the Carrizo segment of the San Andreas Fault (see Figure 6.2), then M8 events will occur on average only every 7,000 years. But by virtue of their enormous size, these events will still spend a significant fraction of the earthquake budget, leaving less for progressively smaller events. Extrapolating back from this M_{max} using a b-value curve, we can predict that M7.5 events will occur only every 700 years.

Recall also that probabilistic seismic hazard analysis typically quantifies the shaking that is expected to occur every few hundred years. If one assumes a model in which M8.0 events happen every 7,000 years and M7.5 events every 700 years, then neither of these events would be considered a sure bet over a period of 500 years. In contrast, if we assume a maximum magnitude of 7.5, then such an event will recur on average about every 400 years. Estimated hazard will be much higher because we would expect the largest event to occur within the time frame considered. We thus arrive at the paradoxical result that a larger maximum magnitude can result in a lower estimated seismic hazard. The bigger the largest event is, the less frequently it will occur.

The debate is by no means restricted to major plate-boundary faults. In the greater Los Angeles region, where complicated systems of smaller faults contribute significantly to seismic hazard, various researchers have argued for maximum magnitudes as low as M6.5–6.7 and as high as M8.5. Given a deformation rate of approximately 1 centimeter per year, we can quantify the frequency of the largest events according to different assumptions regarding their size. Geologist James Dolan and his colleagues showed in 1995 that if the budget is spent only on M6.5–6.7 earthquakes, then such events must occur every few decades. If, in contrast, the distribution includes events as large as M7.5, then events of this magnitude are expected every 300–500 years, with a M6.6 event occurring perhaps every 50–70 years. Such a calculation is an oversimplification because the distribution of events will include a full range of magnitudes up to the largest event (Figure 6.6); however, the calculation is simpler to illustrate when moderate and large events are considered as distinct classes.

A mathematician would regard the question of how to determine maximum earthquake size as an ill-posed one; the question cannot be answered with the available information. Knowledge of fault structure can provide some constraint, but often with a lot of uncertainty. In the end, Earth scientists strive to find what truth they can. Along the San Andreas Fault, the paleoseismic results afford a measure of ground truth. If we assume a maximum magnitude of, say M8.5, and then calculate the rate of M7.8 events to be once every 1,000 years, then we can reject the assumed maximum as too high if there is solid evidence that M7.8 events have occurred every few hundred years.

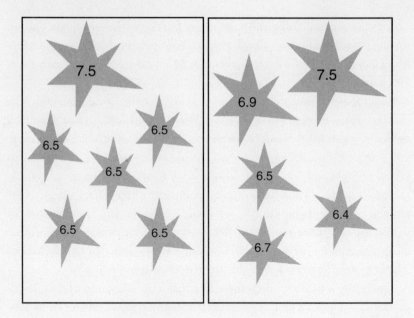

Figure 6.6. Possible distributions of earthquake magnitudes over a given region. Although it is useful to think in terms of a simplified model that has large (M7.5) and moderate (M6.5) events, the distribution shown at right is more realistic.

In the absence of detailed paleoseismic results, we can sometimes look to the historical record for a measure of ground truth. In the greater Los Angeles region, for example, a predicted rate for M6.5 (Northridge-sized) events of one every 20 years is problematic because the observed rate for such events over the last 150 years has been significantly lower. One cannot expect a 150-year record—a mere snippet of geologic time—to be perfectly representative of the long-term average. However, concluding that the historical record grossly underrepresents the average long-term expectations grates against Earth scientists' dearest sensibilities. Such a conclusion implies that the present is somehow unique, which Earth scientists are always loathe to believe.

In the end, determining the expected earthquake distribution for a fault or region requires Earth scientists to finesse questions that cannot be answered by any single equation, experiment, or computer program. The development of understanding is as complicated—sometimes as turbulent but also sometimes as elegant—as the earth itself.

Prediction of long-term earthquake rates in less active parts of the country almost seems easier. Recall Steven Wesnousky's results regarding young faults versus mature faults. In the absence of mature, established fault systems, extrapolations of b-value distributions (from observed small-event rates) may provide a good characterization of the larger-event rate. One doesn't even need to understand the regional fault structure in detail. If the earthquake distribution is described by a b-value curve, then the larger events will happen somewhere.

In intraplate regions—the parts of the crust that are well away from active plate boundaries—maximum magnitude can be estimated from magnitude distributions of similar tectonic settings worldwide. In 1996, seismologist Arch Johnston presented a thorough evaluation of earthquake rates in what he termed stable continental regions, which are areas within intraplate continental crust that are away from failed rifts and other crustal complexities associated with elevated earthquake rates. Stable continental regions are exactly what the name describes—parcels of planet Earth within which deformation rates and earthquake rates are extremely low. Although the historical record of earthquakes is inevitably scant in such places, Johnston argued that one could substitute space for time to constrain maximum magnitude in stable continental regions; that is, by arguing that one would expect similar earthquake distributions in stable continental regions worldwide, we can improve the constraints on maximum magnitude in any one area beyond what can be derived from that region's short historical record.

We should, however, be careful not to gloss over the tremendous uncertainties regarding potential earthquake sources in central and eastern North America. In many ways, the intensity and depth of the debates regarding hazard assessment in California reflect the community's efforts to iron out details now that a gross characterization of hazard has been established. In the absence of information, it is easier to apply the most simplistic formulas and assumptions, but that doesn't mean they are right

In the late 1990s, the depth of our ignorance about earthquake sources in eastern North America was highlighted by an interesting new hypothesis, a proverbial monkey wrench introduced by seismologist John Ebel, who developed a concept known as *paleoseismicity*. Paleoseismicity is not the same thing as *paleoseismology*, the investigation of prehistoric earthquakes on faults. The idea behind paleoseismicity had been floating around for some time. What if aftershock sequences in eastern North America could be so prolonged that a hot spot of present-day seismicity could be viewed as an aftershock sequence rather than as representative background seismicity (recall that both are characterized by the same distribution of magnitudes!). Ebel formalized this concept with mathematics and a rigorous analysis of eastern North America seismicity. He identified eight zones in New England and southeastern Canada where the rate of events appeared to be decreasing over time. By fitting this decrease with the familiar time decay of aftershocks (rate decreasing with the reciprocal of time) and considering the overall rate and distribution of events, Ebel extrapolated back to infer the magnitude and time of inferred prehistoric earthquakes.

Although plagued by nontrivial uncertainties, Ebel's results are compelling. Among the possible prehistoric mainshocks identified are a M7.2+ earthquake that occurred off Cape Ann, Massachusetts, perhaps 2,000 to 4,000 years ago. In 1755, this zone produced a sizeable event whose magnitude has been estimated as M6.5–7. The Cape Ann event is one of the large mainshocks that is known to have occurred in the northeastern United States. But according to Ebel's hypothesis, the event might have been "just an aftershock." Ebel also concluded that the paleoseismicity results give further evidence that a substantial (M6.5–7) earthquake in 1638 was associated with a source zone in central New Hampshire, a region not generally known for earthquake hazard.

The implications of the paleoseismicity hypothesis are sobering. Suppose for the sake of argument that there are twenty source zones in eastern North America capable of generating M7+ events and that these events each occur only every 5,000 to 10,000 years. Now suppose that aftershock sequences of these events can last for a few hundred years. In a historical record only a few hundred years long, what do we observe? The ghosts of mainshocks past, not the harbingers of mainshocks yet to come. The latter could be lurking anywhere, in regions as stubbornly mute as the locked sections of the San Andreas Fault. Although a map derived from present seismicity data would still identify the risk of future moderate (M5–6) aftershocks, the assumption that mainshock hazard is high in regions of great seismicity could thus be completely wrong.

Once again, geology and paleoseismology offer some measure of ground truth—not as much as one might like but at least some. Paleoseismic investigations have revealed a handful of areas to be long-lived loci of large earthquakes. These include Charleston, South Carolina; the New Madrid seismic zone; and the Wabash Valley region in the central United States. However, in intraplate regions where there are few clearly exposed faults, paleoseismic evidence can be gathered only in regions where liquefaction of sediments occurs and where the record of liquefaction will be preserved in the geologic strata over centuries to millennia.

The bottom line for eastern North America is that we don't know what we don't know. The rate of earthquakes is lower there than in the tectonically active West. That much is clear. But the potential for surprise may well be much higher.

PUZZLE PIECE #2: ATTENUATION

Once we have assembled a probabilistic source model for a region, the next task is to predict the ground motions from all sources at all distances. Although a lot of things happen to earthquake waves once they leave the source, one of the most fundamental changes is a decrease in amplitude with distance, that is, attenuation. In hazard mapping, this factor is generally incorporated via attenuation relations, which mathematically describe the level of ground shaking (for a given magnitude) as a function of distance from the source.

And therein lies the first complication: Exactly what do we mean by "level of ground shaking"? We can quantify earthquake shaking in different ways: by looking at how much the ground moves back and forth (displacement), at the speed at which it moves (velocity), or at the rate at which the velocity changes (acceleration). Early attenuation relations were derived from peak acceleration —or peak ground acceleration (pga)—data, which simply involve the peak value of acceleration observed in any earthquake recording. Although using displacement or velocity might seem more natural, strong-motion data have classically been recorded on instruments that record ground acceleration. Early versions of these instruments were analog: they recorded data on film or other nondigital media. Such records could be digitized, but only by a time-consuming process that often resulted in a multiyear delay between the occurrence of a large earthquake and the release of the data. Digitization is computationally intensive and difficult, and without careful technique and execution, the data are easily corrupted.

Peak acceleration data, by contrast, can be read directly from analog records without processing. Expediency therefore begat practice. But although peak acceleration might have seemed to be a reasonable choice for the description of ground motions, seismologists and engineers have come to realize that it is probably not the best choice. The problem with acceleration is that it is controlled by high-frequency ground motions. The M6.9, December 23, 1985, Nahanni earthquake in western Canada is an interesting case in point. After a M6.6 event in the same region on October 5, 1985, the Canadian Geological Survey deployed a number of strong-motion instruments in the region. When the December event occurred, one of these instruments recorded a peak acceleration of more than twice the acceleration of gravity, but as part of a spurious high-frequency burst. Seismologists joked that perhaps the burst had been caused by the kick of a moose that had been startled by the earthquake. A less amusing but more likely alternative is that the burst resulted from rupture of a local asperity that happened to be close to the instrument. In any case, the burst of high-frequency energy, which controlled the peak acceleration, apparently had little to do with the characteristics of the rupture.

Whereas high-frequency motions control peak acceleration, it is the lower-frequency motions that generally damage structures. As a rule of thumb, buildings have their own natural period, which can be estimated as 0.1 second times the number of stories. A two-story house will therefore have a period of 0.2 second (or a resonant frequency of 10 hertz). A ten-story building will have a natural period of approximately 1 second. Seismic waves are dangerous to structures when the frequency of the waves is comparable to that of the buildings (see Figure 6.7 and Chapter 4). A burst of acceleration at frequencies well upwards of 10 Hz will thus be of little consequence for hazard estimation.

For quantifying ground motions with a single parameter, peak ground velocity—the highest velocity of the ground during earthquake shaking—has emerged as the consensus choice because velocity appears to correlate best with damage. However, attenuation relationships have traditionally been based on acceleration because they are derived from data recorded on strong-motion accelerometers.

Recall also the concept of a seismic spectrum: earthquakes radiate energy over a wide and continuous range of frequencies. Attenuation relations would ideally describe how waves of all frequencies are diminished as they move farther from their source. But because developing attenuation relations for a full spectrum of frequencies would be cumbersome, the engineering community has moved toward the derivation of relations for a few discrete frequency points. Three periods are typically used: 0.3, 1, and 3 seconds, a range that generally brackets the ground motions of engineering concern (Sidebar 6.3).

Once parameters for quantification of ground motions are chosen, the next step is to use recorded strong-motion data to determine empirical attenuation

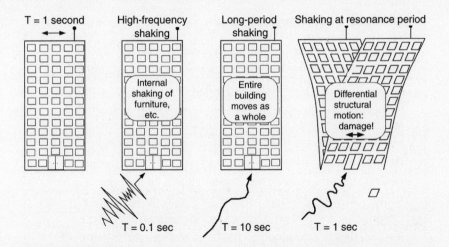

Figure 6.7. The response of a building to input shaking of different frequencies, or periods. If the period of the shaking is short compared to the period at which the building sways *(second panel from left)*, only vibrations internal to the building will result. If the period of the shaking is very long *(third panel)*, the building will move back and forth without significant sway. But if the two periods are close *(right panel)*, then the earthquake waves will shake the structure at its resonance period, and structural damage can result.

relations. Ground motions from earthquakes of a particular magnitude, or a narrow magnitude range, are plotted as a function of distance from the source, and then a smooth curve is fitted through the points. Although straightforward in principle, the analysis is, yet again, fraught with insidious complexity. As a seismologist was heard to quip, fitting a curve through a bunch of data points is not difficult—the art is in extrapolating the curve to magnitudes and distances not constrained by data. Extrapolation is necessary because, unfortunately, the most critical data for hazard assessment—large earthquakes recorded at close distances—are also the sparsest. Because our instrumental record is more meager still than the historical record, the available strong-motion database contains precious few recordings of large earthquakes at close distances.

Derivation of attenuation relations has therefore inevitably entailed some degree of extrapolation. But when simple extrapolations of larger-distance data resulted in implausible levels of near-field ground motions, seismologists and engineers concluded that ground motions must surely flatten at close distances. As the database of large earthquake recordings at close distances began to build,

Sidebar 6.3   Long-Period Energy

Although seismic hazard methodologies do not generally consider ground motions with periods longer than 3–5 seconds, large earthquakes generate substantial energy at even longer periods. Very-long-period ground motions are not considered in standard treatments because little data constrain such motions and because long-period ground motions will affect only the largest buildings and lifelines, such as pipelines. One cannot neglect the hazard posed to large structures by long-period energy, but such structures are not generally designed for a generic treatment of ground motions (deterministic hazard). Instead, they are designed on the basis of a specific assessment of expected ground motions at the site. Tall buildings must also be built with features that allow them to withstand the substantial horizontal forces associated with wind, and these features confer great resistance to earthquake shaking. Finally, tall buildings must be anchored in solid rock, which means they are generally safe from sediment amplification effects.

though, it became apparent that the curves might not flatten so much after all. Theoretically calculated ground motions at close distances appeared to corroborate these observations.

For many years seismologists and engineers faced a real conundrum: Just how extreme could earthquake ground motions be in the near field of a large earthquake? Could velocities as high as 1–2 meters/second, or accelerations well in excess of the acceleration due to gravity, really be achieved? Some lines of evidence indicated that such extreme ground motions were impossible, yet anecdotal evidence—such as the snapping of trees during strong earthquake shaking—suggested that large ground motions did indeed occur.

As evidence of high ground motions recorded at close epicentral distance trickled in, there was a tendency to argue away the data. The high-frequency burst on the one record from the 1985 Nahanni earthquake was arguably spurious. So, too, was the closest recording of the M6.7 1971 Sylmar, California, earthquake—a peak acceleration of $1.7g$ was recorded, but at a station in rugged mountainous terrain. Perhaps the high ground motions represented a local amplification of waves caused by the topographic relief.

The M6.7 Northridge earthquake of 1994 proved to be something of a watershed event because nine strong-motion measurements were recorded on the earth's surface directly, or almost directly, above the dipping fault plane. With so many stations recording high ground motions, it became difficult to argue them away. Instead, the seismological and engineering communities began to pay much closer attention to the issue of extreme near-field ground motions.

But if the community has moved toward accepting high ground motions in the near field of large earthquakes, we still have insufficient data to constrain the attenuation relations as well as we need to. We have little data from the kind of earthquake that most concerns us: a M7.8 rupture on the San Andreas Fault, a M7.5 rupture on the Santa Monica Fault in Los Angeles, a M7.5 rupture in the New Madrid region, a M7.2 event off the shore of Massachusetts.

It is unfortunate that although one can look to geology to provide ground truth for inferences regarding earthquake sources, geology has traditionally provided little information about attenuation relations. In the early 1990s, however, seismologist James Brune proposed that rocks could supply critical information on ground motions from earthquakes that happened in pre-instrumental times—not just any rocks, though, but ones that Brune dubbed "precariously balanced rocks." Observing that certain rock formations weathered in such a way as to create large and precariously balanced boulders (Figure 6.8), Brune argued that these rocks could be used as retroactive strong-motion instruments. If we could estimate the amplitude of ground motions required to topple a given rock and the length of time that rock had been in its precarious state, we could then obtain an upper bound on the ground motions at that location. Because precarious rocks are several thousand years old, this engagingly low-tech hypothesis offered ground motions seismologists something previously almost unimaginable: information on shaking levels from earthquakes that took place long before any seismometers were around to record them.

As Brune and his colleagues began a systematic analysis of precariously balanced rocks throughout California, a general trend began to emerge from their results. Brune concluded that the rocks would have toppled had the ground motions from large prehistoric earthquakes been as large as predicted by attenuation relations and source models. His results suggested, therefore, that the attenuation relations were systematically biased toward predicted ground motions that were too high. This is potentially good news. In at least some ar-

Figure 6.8. A precariously balanced rock near Lancaster, California. Because the rock has existed in its precarious state for some time, we can infer bounds on the maximum level of shaking that could have been experienced at the site over thousands of years. Photograph by James Brune. Used with permission.

eas, the ground might not shake as strongly in future earthquakes as current models predict.

At the time of this writing, however, the seismological community has not yet fully come to grips with the results from studies of precariously balanced rocks. Some scientists have suggested that the rock sites reflect the past ground motions only for point locations where, for whatever reason, shaking from characteristic earthquakes was less severe than usual. Others have argued that the age of the rocks—and therefore the time they have existed in a precarious state—might have been overestimated.

Nonetheless, at a time when earthquake shaking is being investigated with improved methods and instruments, ground motions seismologists are paying attention to the tales that rocks have to tell.

PUZZLE PIECE #3: SITE RESPONSE

Once sources and attenuation are quantified, a final step remains: incorporation of a factor known as site response, by which we mean the effects of local, shallow geologic structure on earthquake waves. The role of near-surface geology in controlling ground motions can be complicated. In this chapter, I focus on the methodology used to distill the complexity into information useful for the construction of hazard maps.

Although the seismological and engineering communities have become convinced of the importance of three-dimensional geometry of basins and valleys in determining site response, inclusion of such complications is beyond the scope of current probabilistic hazard methodologies. In constructing maps, seismologists must fall back on the most elementary method for determining site response corrections: the method involves simple amplification, or sometimes deamplification, factors based either on the near-surface rock type or the seismic shear wave velocity, or both, in the upper 30 meters of ground (Sidebar 6.4). The amplitude of earthquake waves will increase if their velocity decreases, so lower velocities in near-surface layers will translate into amplified shaking

If wave velocities are not directly known, then rock type can serve as a proxy because different types of rock are characterized by different velocities. Seismologists speak of "soft sediments" versus "hard rock," an imprecise taxonomy that differs considerably from a geologist's, which is much more subtle. Yet to a ground motions seismologist, "soft" (low velocity) versus "hard" (high velocity) usually matters far more than any details regarding rock or soil composi-

---

Sidebar 6.4   Why 30 Meters?

The validity of characterizing site type by shear wave velocity in just the upper 30 meters of the crust remains a subject of debate even to this day. Given the near certainty that ground motions are affected by deeper velocity structure, we might ask why 30 meters was chosen in the first place. There is a simple answer to this question. When engineers began to make velocity and other measurements for the purposes of site classification and building design, the most commonly available rigs were built to drill to a maximum depth of 100 feet. Sometimes choices are made for reasons that are less profound than one might imagine!

tion. Soft sediment sites, in particular, are associated with ground motions that are amplified, sometimes greatly so.

Seismologists do consider more than just two rock-type classifications, but given the imperative to keep the methodologies simple, they typically use no more than four: hard rock, weathered rock, old compact sediments, and young unconsolidated sediments. Although it is natural to think of the different site classifications as representing different rock types, modern definitions rely instead on the velocity of shallow shear waves. Although rock type generally correlates with shear wave velocity, the latter more directly controls site response.

The precise nature of sediment-induced amplification depends on how seismic wave velocity varies with depth. Whether sedimentary layers give rise to vibrations that resonate like a violin string or to an amplification that varies smoothly with frequency depends on whether velocity increases abruptly at an interface or rises more uniformly with depth. In either case the net result is the same: the lower the near-surface shear wave velocity, the stronger the amplification effects. Based on large quantities of data, site response amplification factors have been developed for the rock types at the frequencies of engineering interest.

Consider four pans, one each filled with Jell-O, angel food cake, brownies, and burned brownies. If you tap the pans sharply from the side, the jiggling inside the pans will be greatest for the softest material and will diminish as you move to the harder materials. The softest material—the Jell-O—may also oscillate back and forth at one (resonance) frequency. If you were to record the motion inside the pans, you could quantify the amplification of the different materials by employing the multiplicative factors by which jiggling increases as you progress from harder to softer desserts. With these simplified factors, we cannot hope to describe the full range of amplification effects, including the resonances that can greatly amplify waves within a narrow frequency range. Consider the food experiment once again. We might conclude that Jell-O shakes four times as strongly as burned brownies. Yet if we take into account the frequency at which the Jell-O shakes back and forth, the amplification might well be much higher at some frequencies of oscillation. We must either average over all frequencies or develop multiplicative factors for each frequency. In either case, the resulting amplification factors inevitably represent a simplification. They do, however, provide a gross quantification of expected amplification effects in a manner that is appropriate for current hazard map methodologies.

Amplification factors are incorporated as a final step when hazard maps are assembled. First we assess the potential sources (faults) and then calculate how energy leaving each source will attenuate as it travels through the crust. Then we adjust the predicted motions at each site by the appropriate amplification factors.

In California, where active tectonic processes rework the crust continuously and where population centers cluster in sediment-filled valleys and basins, site response corrections are critical to hazard assessment. But site response is an important issue elsewhere in the country as well, given the natural tendency for settlement to cluster along coasts and other waterways. Here again, uncertainties stemming from data limitations loom large. Seismologists have far fewer data with which to quantify expected site response in the central and eastern United States. As population pressures lead to development on marginal land adjacent to bodies of water, a risk evolves that is certainly real but not well understood.

For decades, application of site-response corrections (anywhere) was plagued by a lingering uncertainty—or perhaps, more fairly, a lingering controversy—concerning the linearity of ground motions. Site response is said to be linear if the amplification caused by near-surface sediments is independent of ground motion amplitude. If the ground motions that enter a sediment-filled valley are scaled up by a certain factor, the ground motions at the surface will scale up by that same factor. (Tap the four dessert pans twice as hard, and the relative amplitudes of jiggling will remain the same.) With the alternative model— a nonlinear sediment response—the amplification becomes progressively smaller as ground motions intensify. The dominant frequency of shaking also shifts nonlinearly. Nonlinear behavior occurs because unconsolidated sediments lose their internal cohesion when subjected to extremely strong ground motions.

By the 1980s the engineering community had long since concluded that unconsolidated sediments would exhibit substantial nonlinearity; this conclusion was based on laboratory experiments in which crustal materials were subjected to high strains. The seismological community, however, remained skeptical, having failed to find compelling evidence of nonlinearity in earthquake data. As late as the early 1990s, earthquake scientists remained unconvinced that nonlinearity was significant.

The seismological community might appear, in retrospect, to have been too dismissive of results from a different discipline, since they had so few recordings of strong ground motions with which to address the issue directly. But part of the skepticism resulted from the engineers' interpretation that a loss of cohesion (a nonlinear response) would impose a stringent upper limit on peak ground motions at sediment sites. Engineers estimated upper bounds as low as 0.3g. Even in the absence of abundant near-field instrumental data, seismologists had many documented accounts of earthquake effects (large objects being thrown into the air, for example) that belied a bound as low as this.

In any case, convincing observational evidence for a nonlinear response began to trickle in as seismologists built their database of moderate to large earthquakes recorded at close distances. While the data dispelled the notion that ground motions at sediment sites were sharply limited, they did begin to provide evidence for a nonlinear site response. In 1997, seismologist Ned Field analyzed data from the Northridge earthquake and presented one of the first compelling cases for pervasive nonlinear response. By comparing ground motions from the M6.7 mainshock with motions from aftershocks recorded at the same sites, Field concluded that the strong-motion site response differed significantly from the weak-motion response. The nature of the difference—lower amplification and a suggested shift toward lower-frequency motions—was consistent with theoretical predictions.

Nonlinearity can be good news; it means that amplifications of the strongest ground motions will be less severe than predicted from a linear model. Suppose that when you bang the pan of Jell-O hard, its jiggling is no longer larger than that of the other pans by the same multiplicative factor. Given the nature of nonlinearity—in particular the loss of cohesion of a solid material—the factor will be smaller. Thus, for large earthquakes, the amplification at soft rock sites could be less severe than for smaller earthquakes.

For the purposes of hazard assessment, however, nonlinearity is a monkey wrench in the works because it means that the substantial body of site response results derived from small earthquakes might not be applicable for the prediction of shaking from large events. As seismologists and engineers strive to improve our understanding of site response for future generations of hazard maps, the issue of nonlinearity will likely remain a hot area of research. Much work has already been done on the final piece in the hazard mapping puzzle. And with all of the requisite pieces assembled, the final construction can begin. Much work remains.

PUTTING THE PIECES TOGETHER

We can now examine how the pieces are put together to create probabilistic seismic hazard maps. In the 1990s, seismologist Art Frankel and a team of scientists from the U.S. Geological Survey led a national effort to produce a new generation of hazard maps for the United States. In California, Frankel collaborated with a team of seismologists from the California Division of Mines and Geology. Rather than tackle site response in the first iteration, Frankel and his colleagues mapped a site condition that was considered average in some sense. Examples of these maps are shown in Figure 6.9 for the United States as a whole and for California.

Almost as soon as the maps were complete, government seismologists started to lay the groundwork for the next generation of hazard maps. The new maps will incorporate site response terms, possibly including nonlinearity. Maps will also incorporate any revisions to source models and attenuation relations deemed appropriate since the last iteration. We must undertake this revision process, expected to continue into the foreseeable future as knowledge grows, with great care to maintain a balance between applying our knowledge and confessing our ignorance. Hazard maps represent consensus products; they draw on results that bear the imprimatur of the scientific community. Inevitably such applied methodologies are often one step behind cutting-edge scientific research. The long, sometimes slow process whereby research is translated into practical applications can be a frustrating one, but the alternative—basing hazard maps on research not yet definitively established—is untenable.

DETERMINISTIC HAZARD

Although hazard mapping was a central activity of Earth scientists through the 1980s and 1990s, it was not been the only one. Parallel efforts to understand so-called deterministic seismic hazard have continued apace. A deterministic approach seeks to quantify not the hazard that is expected in a region but the hazard that would result from the rupture of a given fault segment. Deterministic hazard assessment is useful primarily for building design. The ground motions that will shake a building are not the average motions from all sources, but the actual ground motions from one source.

Peak Acceleration (%g) with 10% Probability of Exceedance in 50 Years

site: NEHRP B-C boundary

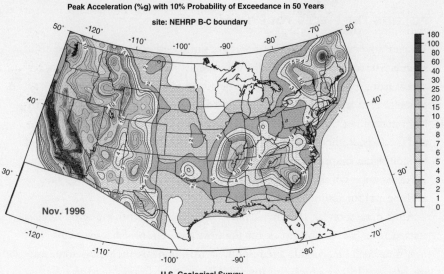

Nov. 1996

U.S. Geological Survey
National Seismic Hazard Mapping Project

Peak Acceleration (%g) with 10% Probability of Exceedance in 50 Years

site: NEHRP B-C boundary

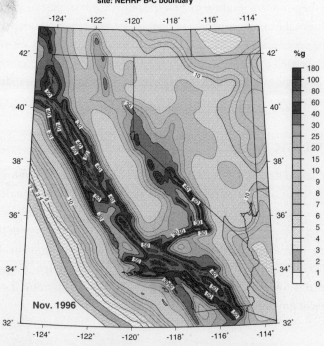

Nov. 1996

For California portion: U.S. Geological Survey - California Divison of Mines and Geology

For Nevada and surrounding states: USGS

Recent advances in our understanding of earthquake source processes and of the propagation of earthquake waves are relevant to deterministic hazard assessment. Through the 1990s, one area of emphasis in the seismological community was scenario earthquake modeling: the calculation of expected ground motions at a grid of surface sites, motions that result from the rupture of a fault segment with specified parameters. Typically, the rupture dimensions and average rupture velocity are applied by way of one of several schemes.

The work of Kim Olsen, who investigated the response of large basins and valleys to earthquakes on different faults, is an example of ground motions prediction for scenario earthquakes. Olsen's conclusions regarding the variability of basin responses (they depend on the earthquake's source location and the details of the rupture) illustrate the daunting challenges of the task. In a simple world, earthquake shaking would depend only on earthquake magnitude and on distance from the fault, but the real world is far from simple. Olsen's results tell us that a realistic prediction of shaking from individual fault segments requires an understanding of earthquake sources as well as a grasp of wave propagation in the inevitably complex, three-dimensional earth.

By the mid-1990s, at least a half dozen approaches to the prediction of ground motions from scenario earthquakes had been developed. Different theoretical and numerical methods rely on different computational approaches; other methods are empirical rather than theoretical—that is, they are based on empirical quantification of past earthquakes rather than on equations derived from physical principles. Arriving at consensus in scenario earthquake model-

Figure 6.9. *Top,* hazard map for the entire United States as determined by the U.S. Geological Survey. Shading indicates the level of shaking (expressed as percentage of acceleration of gravity) that is considered to have a 10 percent probability of occurrence over a 50-year window. The active western plate-boundary system gives rise to the highest hazard in the contiguous United States; away from the West, zones of high hazard are inferred at the New Madrid region, in coastal South Carolina, and along the Saint Lawrence Seaway (eastern Canada). *Bottom,* hazard map for California, by the U.S. Geological Survey and the California Division of Mines and Geology. This map shows shaking estimated to have a 10 percent probability over 50 years. (Original maps were generated using a color scale, which results in some ambiguity when the maps are printed in black and white. Color maps and extensive documentation are available on line.[19])

ing is quite a bit more complicated than arriving at consensus with probabilistic hazard assessment. Predicted ground motions from a given earthquake at a given site are not amenable to simple averaging schemes. Yet seismologists will continue to test available modeling methods against real data and against each other. With the computer technology and theoretical understanding already in hand, advancement is virtually assured. And the better we understand the hazard, the more we can do to mitigate the risk.

In the past, the business of designing earthquake-resistant buildings has all too often brought into play the exercise of learning from past mistakes. Engineers have sometimes been surprised by the failure of buildings thought to be earthquake resistant. For example, unexpected damage was found in the welds of steel-frame buildings following the 1994 Northridge earthquake. Although building design is the purview of engineers rather than scientists, the seismologist plays a critical role by specifying the nature of the expected ground motions. Thus, as scenario earthquake modeling matures to the point where realistic ground motions can be predicted, we will, for the first time, be able to use computer simulations of building response to determine what will happen to a building when it is subjected to strong ground motions—before the building falls down.

REMAINING ISSUES

Even as the seismological community strives to apply its current body of knowledge to the goal of seismic hazard assessment (and, ultimately, risk mitigation), one eye always remains focused farther down the road. In the arena of hazard assessment, it is always imperative to apply our knowledge and to work toward remedying our ignorance.

After a century or so of instrumental seismology, a handful of unresolved issues remain the focus of ground motion research efforts. Can we go beyond determining the average expected hazard in a given region and begin to quantify the hazard as a function of time (the time-dependent hazard). How do we incorporate three-dimensional basin effects into our quantification of probabilistic hazard? Can earthquake shaking depend significantly on small-scale details of the rupture? How strong are the ground motions in the near field of large events? How important is the nonlinear response of sediments?

Research efforts continue apace on these and other unresolved issues. In some cases it is not yet clear how present-day research will translate into hazard application, but past history and common sense suggest that an improved understanding of ground motions cannot help but improve our ability to quantify hazard.

Taking as an example just one of these questions, the issue of time-dependent hazard, we find that progress appears tantalizingly close. The critical point model suggests that moderate earthquakes offer clues about the state of stress in the crust and, therefore, possible clues about impending large earthquakes. And recently developed theories regarding earthquake interactions suggest that given the earthquakes that have already happened, we might be able to predict where earthquakes are more—and less—likely in the future. Both of these theories therefore promise to be useful for refining our time-dependent hazard estimates according to how much we know about the current state of the earth's crust. And maybe, just maybe, both hold some hope for intermediate-term earthquake prediction.

Even at the Earth science community's most pessimistic point regarding short-term prediction, some scientists held out hope that progress might be made on intermediate-term prediction—that is, the identification of fault segments ripe for large earthquakes in the coming years or decades. The Parkfield Prediction Experiment was an example of intermediate-term prediction.

Another notable intermediate-term prediction effort was carried out by a group of experts in the late 1980s charged with quantifying hazard along the San Andreas Fault. Dubbed the Working Group on California Earthquake Probabilities (WGCEP), the group drew on its collective knowledge of the earthquake history of the San Andreas: when the major segments last ruptured (Sidebar 6.5), information on slip patterns in the most recent earthquakes (the 1906 San Francisco earthquake especially), and their general understanding of earthquake cycles.

The WGCEP's results were summarized in a single figure (see Figure 6.10) that depicted the identified segments of the San Andreas Fault along with their assigned probability of producing an earthquake over the next 30 years (1988–2018). When the 1989 Loma Prieta earthquake occurred along a segment that had been identified as the second most likely (after Parkfield) to rupture, seismologists rushed to point to the results presented in Figure 6.10, which seemed to be a highly successful intermediate-term earthquake prediction.

The working group, however, had scarcely been unanimous on the identification of the Loma Prieta segment as particularly ripe for a large event. Indeed, individual group members had argued for dramatically different interpreta-

Sidebar 6.5   Seismic Gap Theory

By combining the theory of seismic strain accumulation with the characteristic earthquake model, we arrive at the most fundamental formulation for long-term earthquake forecasting: *seismic gap theory.* A seismic gap is defined as a segment of a major fault that, by virtue of its lack of seismicity over a certain time period, is assumed to be locked. With our knowledge of plate motions and the size of earthquakes produced by a plate boundary—be it a transform fault or a subduction zone—we can estimate the average rate of great earthquakes on that boundary. If we also know the earthquake history of a fault, we might reasonably conclude that intermediate-term earthquake prediction could be based on seismic gap theory. Through the 1980s and 1990s, tests using the global earthquake catalog failed to unambiguously support this idea. Although seismic gap theory must have some element of truth, the utility of the theory for earthquake prediction may inevitably be limited by complications such as stress interactions.

tions of available data. On the basis of the low slip inferred for this segment in the 1906 earthquake, some maintained that rupture on this segment had a high probability; others disagreed and favored a low probability because a major earthquake had ruptured this segment as recently as 1906. The consensus probability that was eventually assigned—high but not extremely so—was very much a compromise.

Perhaps it is fair to conclude that the WGCEP assignments were an eminently successful consensus. Perhaps not. This cup may be destined to remain either half empty or half full, depending on one's philosophical point of view. Yet the WGCEP's efforts have clearly been a tremendous learning experience, one that allows the community to learn from its mistakes, as well as its successes, and move forward.

The probabilities assigned by the WGCEP were based mainly on the most basic tenets of seismic gap theory: the longer it's been since the last earthquake, the sooner the next one will strike. But in the decade since this landmark effort, the community has developed new insight into the nature of the earthquake cycle. At least according to current observations, an increase in the regional rate of moderate events may herald large earthquakes. If further re-

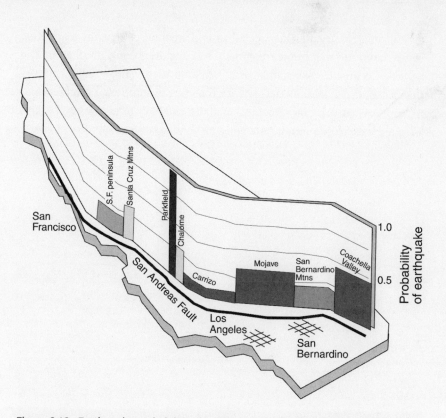

Figure 6.10. Earthquake probabilities for different segments of the San Andreas Fault determined by the Working Group on California Earthquake Probabilities. The height of each bar corresponds to the determined probability of an earthquake between 1988 and 2018. The shading of each bar indicates the estimated reliability of each probability, with the darkest shading indicating the highest reliability. Adapted from WGCEP results.[20]

search substantiates these early results, and if an effective scheme can be devised to identify, a priori, the extent of regions approaching criticality, then forecasting heightened regional probability, over a decade or so, may become feasible.

In a similar fashion, monitoring of stress changes caused by previous large earthquakes could help to refine long-term hazard assessments. In this instance, more information could be used to assign higher and lower earthquake probabilities, depending on the nature of the stress change in different regions. But Ruth Harris's conclusions regarding earthquakes that pop up in the stress shadows of previous earthquakes (see Chapter 3) remind us that there are often complications to the simplest theories. Her conclusions raise another red flag as well. Confucius' exhortation to humankind is sometimes more complex than it sounds. Sometimes the real trick is neither to apply what you know nor to confess what you don't know. Sometimes you have to take a step back to ask a more difficult question: Are you sure you know what you don't know?

SEVEN A JOURNEY BACK IN TIME

We live between the mists of yesterday and the clouds of tomorrow.
Rarely are we given a chance to step outside of time.

—FROM A COWICHAN LEGEND

A seismologist walking along a deserted beach comes upon a curiously shiny object. He picks up the object and discovers it to be a vintage oil lamp. As he starts to dust it off, he is startled by the sudden appearance of a genie. The genie thanks the seismologist profusely for liberating him and, being a specialized sort of mythical guy, announces that he will grant three earthquake-related wishes. What might our fortunate seismologist wish for? The first wish would be easy: the ability to predict the time and place of large earthquakes. His second wish might be for more knowledge of earthquakes that happened tens of thousands of years ago. The third wish, however, might give our seismologist pause. He might wish for insight into a problem related to his own research, or perhaps he might wish for an understanding of earthquake ground motions sufficient to enable him to predict how the ground would shake during an earthquake. An unusually ambitious wish-maker might even ask for the ability to prevent earthquakes. While undoubtedly arising from noble sentiment, this wish could lead to a dark science fiction tale in which an Earth that supports no earthquakes might also (one could imagine an author of fiendish imagination writing) support no life.

But let us return to the second wish, which stems from a frustration shared by virtually all Earth scientists. We strive to understand the earth's seismicity given the minute sliver of time for which we have data. Even the longer historical record, which provides observations of large earthquakes but no recorded seismic data, is the barest blink of an eyelash against the grandeur of the geologic timescale. It is the stuff that fairy tales are made of, the wish to extend our knowledge back in time.

A SHORT-TERM GLIMPSE OF LONG-TERM PROCESSES

Are earthquakes periodic or random in time? Do events strike in clusters that consist of multiple events over short time intervals and are followed by much longer intervals of low seismicity? Are there patterns of seismicity that herald the arrival of a large earthquake? How does the wallop of one large earthquake affect the occurrence in time and in space of future large events? These are just a few of the questions that vex seismologists—primarily because of our limited record of past earthquakes—questions that, as I pointed out in the last chapter, sometimes have important implications.

In the most seismically active plate-boundary regions, a few hundred years can just begin to provide a view of the earthquake cycle—the repeated progression whereby a fault is loaded (stressed) by slow plate motion, releases the stored stress in an earthquake, and begins the process all over again. We know that these cycles are not as regular as clockwork. Earthquake recurrence intervals on a segment of fault can vary by at least a factor of two. Even so, the general concept of an earthquake cycle is still considered valid. A major plate-boundary fault being stressed at a rapid rate will suffer earthquakes much more frequently on average than a fault subjected to a lower rate of stress.

On the San Andreas Fault, the earthquake cycle is a dance that plays out over perhaps 150 to 350 years, the average recurrence interval for large (M7.8–8) earthquakes on each of the three major segments of the San Andreas (Figure 7.1). This fault is thought to be incapable of rupturing from end to end, most notably because the middle section creeps along at a steady rate without ever storing enough stress to generate large earthquakes. A rupture that starts on one side of the creeping section is unlikely to be able to blast through a region where stress never accumulates. It would be a little like trying to crisply tear a creased newspaper and suddenly encountering a soggy patch; the energy of the tear would dissipate and its forward progress would stop.

The northern segment of the San Andreas Fault is bounded on the south by the creeping zone and on the north by the intersection of three plates, where the fault comes to an end. The nature of the fault segment that runs from the southern edge of the creeping section down to its southern terminus near the California-Mexico border is less clear. The northern part of this 500-kilometer segment of fault is known to have ruptured in the M7.9 Fort Tejon earthquake of 1857. The southern segment appears to have last ruptured in the late 1600s. Although previous earthquakes have not respected consistent segmentation, it

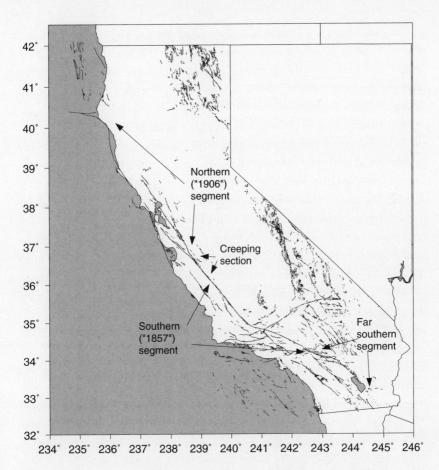

Figure 7.1. The principal segments of the San Andreas Fault in California. These segments are sometimes broken down further, as illustrated in Figure 6.10.

is reasonable to think in terms of two major segments to the south of the creeping zone: the 1857 rupture zone and a southernmost segment stretching from San Gorgonio Pass just east of San Bernardino down to the Salton Sea.

In a historical record that is barely 200 years long, it is not surprising that we have written accounts of large earthquakes on only two of the three major fault segments: the 1857 Fort Tejon event and the great San Francisco earthquake of 1906. We have no historical evidence of the most recent event on the southernmost segment, although its timing has been inferred from geologic evidence that I explore later in this chapter.

For the contiguous United States, this is as good as it gets. The historical record in the Pacific Northwest is comparable to that in California, but the average recurrence rate of large earthquakes appears to be more like 300–500 years in the former region. In the east, we are blessed with a historical record that is about twice as long as that of the West, but that record is cursed by earthquakes that play out over timescales perhaps a factor of ten longer.

As if all of this weren't bad enough, there is another complication. To fully quantify and understand earthquakes, we need recordings of the earthquake waves, but our instrumental record spans less time still than our historical record. We have virtually no seismograms for events prior to 1900, and instrumental coverage remained sparse for many decades after the turn of the century. So we are left with conundrums such as the great New Madrid earthquake sequence of 1811–1812; we know that three huge earthquakes hit in an unlikely spot, but their sizes remain uncertain to nearly a full magnitude unit.

Knowledge of regional earthquake history is critical for quantifying future long-term earthquake hazard. In this chapter, I focus on the ingenious methods that Earth scientists have devised to go one step better than quasi-blind extrapolations of currently observed seismicity rates; what is important here are the methods that extend our direct knowledge of past earthquakes beyond the instrumental and historical records. How have we established dates for past earthquakes for which there are no written reports? How have we unraveled the details of historical earthquake ruptures for which we have no seismic data? These are fascinating questions, and they have equally fascinating answers.

THE PACIFIC NORTHWEST: ANCIENT TALES, MODERN SCIENCE

Since the earliest days of plate tectonics theory, it has been recognized that the Pacific Northwest sits atop the Cascadia subduction zone, along which the oceanic crust offshore is sinking beneath the edge of the continental crust delineated by the coastline (Figure 7.2). Most subduction zones worldwide can create large earthquakes. A few, such as the Mariana subduction zone in the South Pacific, are largely aseismic; tectonic forces are mostly accommodated by steady, slow motion (and small earthquakes), as in the creeping section of the San Andreas Fault. Such subduction zones are considered weakly coupled.

Strongly coupled subduction zones include many regions along the Ring of Fire that are well known for high earthquake hazards. These zones include the western coast of South America, which in 1960 produced one of the largest

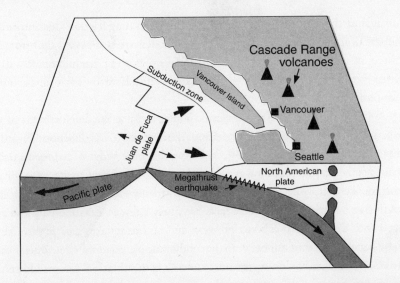

Figure 7.2. The Cascadia subduction zone, where the Juan de Fuca oceanic plate subducts beneath the North American plate.

earthquakes (M9.5) ever recorded; the Alaska coast, which produced the second largest recorded event, the devastating M9.2 Good Friday earthquake of 1964; and the subduction zone under the island nation of Japan, which produced the great Kanto earthquake (M7.9) of 1923.

The maximum magnitude of an earthquake at a subduction zone is primarily a function of two factors. The first, the rate at which one plate is subducting under the other, is fairly intuitive; rapid rates of subduction lead to well-developed fault zones that can produce huge fault ruptures. The other factor, the age of the oceanic crust being subducted, is a little more complex. Oceanic crust is generated by the continuous upwelling of magma at mid-ocean spreading centers at a rate of 2–7 centimeters per year. New oceanic crust is warm and buoyant; as it ages and is pushed farther away from the ridge, it cools and becomes less buoyant. Only a matter of geometrical happenstance determines how long oceanic crust exists before being recycled back into the mantle at a subduction zone. If the spreading center is close to a continent, the time can be as short as 10 million years. If the spreading center is half an ocean away from a subduction zone, the oceanic crust can have a lifetime upwards of 100 million years. Being less buoyant, older and cooler crust subducts more easily than does younger crust. Young crust, meanwhile, is a little like cork bobbing up against the continental crust and thereby generating friction that impedes subduction. With less resistance to overcome, less stress is stored up before a parcel of old crust subducts in an earthquake. Thus the subduction of older crust is associated with smaller earthquake magnitudes than is the subduction of younger crust.

As late as the early 1980s, the seismic potential of the Pacific Northwest subduction zone was still a matter of debate. Although there was dissension regarding the influence of several factors, including the thick wedge of sediments that blankets the oceanic plate, it was easy to think the zone would create no massive earthquakes simply because it had not generated any such earthquakes to speak of during historical times. There have been only a few moderate events associated with the Cascadia subduction zone, and no earthquakes greater than M4 have occurred on the fault between the subducting plate and the overriding crust (the so-called subduction megathrust fault). It is this interface that would generate any massive subduction earthquakes. Compared to the seismicity levels elsewhere along the Ring of Fire, this pronounced quiescence is extraordinary.

In the early 1980s, seismologists Tom Heaton and Hiroo Kanamori carefully compared the Cascadia subduction zone to other subduction zones worldwide. They compiled compelling evidence that the zone was indeed capable of producing earthquakes of at least M8.5. Their conclusion was based primarily on a subduction rate that was low but not unprecedented for subduction zones known to have produced large earthquakes, and on the exceptionally young age of the submerging oceanic crust.

In light of this new assessment, the striking absence of large earthquakes off the Washington and Oregon coasts took on a sinister rather than a comforting air. A stone-quiet major fault zone is like a coiled snake—motionless but capable of striking with a vengeance. The locked segments of the San Andreas Fault are well-known examples of this type of snake, and by the mid-1980s, most Earth scientists had concluded that the Pacific Northwest megathrust was another.

But where was the smoking gun? Given the paucity of events in the historical record, there was little direct evidence for prehistoric Pacific Northwest events. Several scientists noted that Native American legends told of great earth shaking and ocean disturbances evocative of major tsunamis. One legend told the tale of a great earthquake that had struck on a winter night long ago.

Although such legends are certainly captivating, careful examinations of the oral traditions of native peoples have revealed that the legends are fraught with inconsistencies. In 1868, for example, legends concerning events just 50 years earlier were found to vary widely from individual to individual. Seismologists are generally skeptical of even direct subjective accounts of earthquake shaking (Sidebar 7.1). Perhaps unfairly, seismologists discount the ability of nonspecialists to report accurately on the nature of ground motions during the intensely stressful circumstances presented by a major earthquake. For example, a nearly universal assumption is that the reported duration of shaking will be far longer than the actual duration. And any inaccuracies created by our subjective perceptions can be enormously magnified as they are passed down through the generations.

Between 1980 and 1990, hard evidence began to emerge for large prehistoric events that had altered the coastline of Washington and Oregon. Geologist Brian Atwater found regions along the Washington coast where marsh and forest soils had been buried abruptly by land-directed surges of salty water. His observations were consistent with the ground motions expected for a large subduction earthquake followed by a tsunami. By the early 1990s, Atwater had uncovered strong evidence for two events: one approximately 300 years ago, and one 1400–1900 years ago. Less conclusive evidence suggested a third event of intermediate age.

The real smoking gun was not unearthed until the mid-1990s, however, and in the most unlikely place. Digging into ancient Japanese tide-gauge records, seismologist Kenji Satake and his colleagues focused on evidence of several large tsunamis that were not associated with large earthquakes in Japan. Orphan tsunamis—those that occur without accompanying earthquakes—had been recognized for many years. Satake realized that the timing of one particular orphan tsunami coincided with the date of one of Atwater's inferred events, January 26, 1700. Satake analyzed the records and showed them to be consistent with only four possible earthquake ruptures along the Pacific Rim. Two of these regions had been settled by the year 1700: Kamchatka (settled by Russian immigrants beginning in the 1680s) and southern South America (where we have written earthquake reports from as early as 687). There was no record of a massive earthquake in the year 1700 in either place. A third region, Alaska, was not settled, and there was no geologic or historical evidence for an earthquake large enough to generate the tsunami in Japan (that is, M9.4). An earthquake that massive almost certainly could not have occurred without leaving some geologic trace.

That left the last possible source region: the Cascadia subduction zone, where the dates inferred from the earlier geologic investigations were consis-

Sidebar 7.1 Bearing Witness

Although the limitations of subjective reports of earthquake shaking are evident, some individuals do retain a remarkable clarity of thought under extremely stressful conditions. One such individual was living in the small town of Yucca Valley, California, when the 1992 Landers earthquake struck just a handful of kilometers to the north. Although the Landers mainshock ruptured north from its epicenter, two small tails of surface rupture were mapped to the south of the epicenter on separate faults. One of these faults (the Eureka Peak Fault) ran right through the house where a 15-year-old boy and his family were living. When geologist Jerome Triemann arrived to map the fault, the young man told his story of that fateful morning. Awakened by the Landers mainshock, he had rushed outside and watched the fault rupture through his back yard 20 seconds later.

This report left me perplexed because I had analyzed the seismic data and felt confident that the Eureka Peak Fault had ruptured in a large aftershock approximately 3 minutes after the Landers mainshock. The eyewitness account didn't seem to make sense: usually, observers are convinced that the shaking lasted longer than it really did. Eventually I took a closer look at data recorded from the Landers mainshock itself and discovered something curious. One set of wiggles late in the record appeared to have arrived from a different direction than the waves earlier in the seismogram. After much careful analysis, I concluded that these wiggles were consistent with a rupture on the Eureka Peak Fault, and that the fault therefore must have ruptured twice within minutes (an unusual but not unprecedented occurrence). These wiggles showed up on the seismograms about 40 seconds after the initial energy from the mainshock arrived. It was just enough time to let the young man wake up and rush outside with about 20 seconds to spare.

tent with the timing of the 1700 tsunami. And in a coup de grâce that raises goose bumps, the inferred local time of the January 26, 1700, event was 9:00 P.M., long ago on a winter night. It was a massive earthquake that forever re-arranged the coastline of the Pacific Northwest and shook the planet hard enough to generate significant waves on the other side of the Pacific Ocean.

NEW MADRID: A RIDDLE INSIDE AN ENIGMA

The great earthquake sequence that rocked the southeastern United States in the early 1800s was named for a small town that cannot be found in any current travel guide. Within the Earth science lexicon, the town of New Madrid shares a distinction with the tiny town of Parkfield, California, in that both names are instantly and universally recognized by Earth scientists but are otherwise almost entirely obscure.

The largest modern city in proximity to the New Madrid seismic zone is Memphis, Tennessee, the birthplace of the blues. Memphis is a city with a soul, but in a geologic sense it is a city with an Achilles heel. Memphis is located on the Mississippi embayment, atop a crustal structure known as the Reelfoot Rift (Figure 7.3).

Away from plate boundaries, continental crust is stable compared to oceanic crust, which is created and recycled every 10–100 million years. But continents are scarcely inviolate; they can be altered in several ways by the inexorable forces of plate tectonics. Sometimes they are even torn apart. It is difficult to imagine an entire continent being ripped apart. This process does not occur the way it is depicted in bad Hollywood movies; individual earthquakes do not pull the earth apart, leaving behind great chasms. Instead, the slow and steady motion of normal faults accommodates extension and creates a rift zone in which the crust is stretched progressively thinner.

If anything on Earth is insistent, tectonic forces are. The activity related to mantle convection and plate tectonics gives rise to plate motions that continue for millions of years. Yet the limited spherical geometry of our planet introduces what mathematicians call boundary conditions: a plate cannot stay fixed in its motion forever because it ultimately encounters complications caused by other plates that are doing their own thing. Plate motions remain constant for millions of years, but not forever.

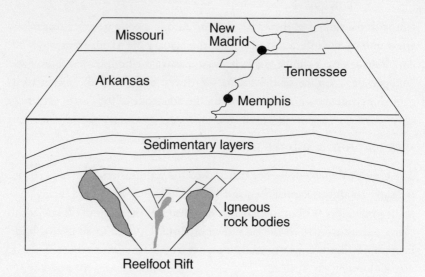

Missouri

New Madrid

Arkansas

Tennessee

Memphis

Sedimentary layers

Igneous
rock bodies

Reelfoot Rift

Figure 7.3. The Reelfoot Rift in the New Madrid region. A process of extension, or rifting, is inferred to have gone on up until about 500 million years ago. This process stopped, leaving a failed rift that is thought to correspond to a zone of relative weakness in the midst of the otherwise old and strong continental crust.

Sometimes the forces creating a rift zone come to a stop and leave what geologists term a *failed rift*. The Reelfoot Rift represents the scar of one such aborted process. Now deeply buried under the massive blanket of sediments carried down over the ages by the Mississippi River, the Reelfoot Rift can be thought of as a zone of weakness slicing through a continental crust that is otherwise very thick, very old, and very strong.

We should not carry the analogy too far, though. Just like a fault at depth, a zone of weakness deep in the crust is still strong beyond imagination; it is weak only in relation to its surroundings, but sometimes even this proves to be critical. To understand the importance of the Reelfoot Rift's being a zone of weakness, we have to consider the broad tectonic forces that affect eastern North America. While not ravaged by the forces that affect an active plate boundary, the eastern side of the continent is subjected to a slow but steady force of compression. The compression comes from the mid-Atlantic spreading center, which pushes everything west of the Mid-Atlantic Ridge farther westward. Because crustal material is not like a sponge—it cannot simply become more dense when compressed—compressional plate tectonic forces will, inevitably, thicken the crust.

Modern earthquakes recorded in eastern North America are thought to be largely due to the effect of two forces: the east—west compression caused by mid-Atlantic spreading and, to the north, an upward force associated with postglacial rebound, the slow upward motion, or rebound, of the crust after having been pushed downward by massive ice sheets during an ice age.

Although it is unaffected by postglacial forces, the New Madrid region is subject to the same broad east—west compression as the rest of the continent. Computer models have been developed to show how tectonic stresses are concentrated at the Reelfoot Rift, with its deep geologic properties and its geometry. If the vast stretch of the earth's crust from the Mid-Atlantic Ridge to the western plate-boundary system is thought of as a chain, the failed rift is one of its weakest links. Pull or push for long enough and the crust will break at the point of weakness.

Unlike the Cascadia subduction zone, the New Madrid zone has not been the subject of debate regarding its capacity to generate large earthquakes. The earth itself settled the issue early in this country's history, with a vengeance. In Chapter 3, I wrote about the New Madrid sequence in the context of earthquake mechanisms. I return to it now to focus on the issue of historic and prehistoric events.

Earth scientists are fortunate to have historical accounts of the most recent New Madrid sequence. These accounts provide far greater detail than do geologic data alone. Like the 1700 Cascadia event, the first of the great New Madrid earthquakes shattered the stillness of a winter night. It occurred in the wee hours of the morning of December 16 in the year 1811, on a clear and quiet night that had given no hint that a seismological event of massive proportions was about to commence. Perhaps there were foreshocks on that fateful night, too small to be felt by the region's sleeping inhabitants; we will never know. The first temblor to be felt struck at approximately 2:15 A.M., and it was a tremendous show of power.

A sizeable aftershock struck approximately 30 minutes later, and written accounts document the occurrence of at least two dozen additional felt earthquakes before dawn. At first daylight, a second powerful earthquake struck, reported by some to be as powerful as the first event.

Over the next five weeks, dozens of events were faithfully and scrupulously recorded by a number of individuals. One of the most complete records came from a Mr. Jared Brooks of Louisville. Over the course of three months, he chronicled 1,874 separate events and classified them by size. The magnitudes of the smallest events, described as only barely felt, were likely close to M3. The size of the largest events, described as "most tremendous, so as to threaten the destruction of the town," remains a subject of debate to this day.[21]

The rate of the strongest aftershocks apparently continued to diminish in January until the morning of January 23, when the region was rocked once again by an event widely described as comparable in intensity to the first. This event was followed by another burst of aftershocks ranging from "barely perceptible" to "considerable shock." Finally, in the wee hours of the cold morning of February 7, the great sequence was punctuated by a fourth large event—the "hard shock" widely described as the strongest temblor of all.

Written accounts of this incredible sequence have been scrutinized and evaluated with painstaking care by many individuals. Although the observations are largely subjective and therefore involve uncertainties we can barely even guess at, some information can be gleaned with confidence. These include the distances from the New Madrid region over which the events were felt (Figure 7.4), a general sense of the shaking level from descriptions of damage to structures, and the strengths of different events reported at different locations. Descriptions of earthquake effects, including damage to structures and perceptions of severity, can be used to assign Modified Mercalli Intensity values. This scale, developed long before instrumental recordings of earthquakes were available, ranges from values of II–III for barely perceptible shaking to XII for shaking that causes near-complete destruction.

The extent of liquefaction, documented historically and largely still observable today, further constrains the magnitude of the shocks. Liquefaction is a phenomenon whereby underground saturated sand layers shake so violently that they lose their coherence as a solid and suddenly begin to behave like liquids. If a sand layer is capped by a layer of compacted material such as clay, the pressure in the liquefied sand layer can rise enormously.

One type of liquefaction feature is known as a *sand blow*. A sand blow occurs when the rising pressure at depth forces sand up narrow conduits through the overlying layers until it literally blows through the surface. The signature of these fountains is a circular deposit of sand left on the surface once the shaking has stopped. The extent of the sand blows resulting from the 1811–1812 earthquakes was documented by Myron Fuller in 1905. A region of extensive sand blow features stretched approximately 130 kilometers in length and

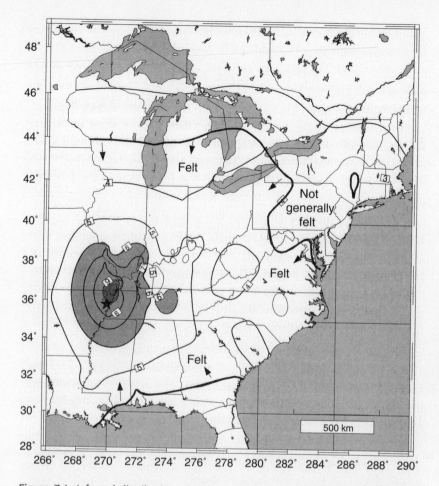

Figure 7.4. Inferred distribution of shaking for the first large New Madrid earthquake, which occurred at approximately 2:15 A.M. on December 16, 1811. From firsthand accounts of the shaking, seismologists have determined Modified Mercalli Intensity values, which range (here) from III (strong enough to be felt but not generally strong enough to wake people up) up to X (extreme damage). The dark line indicates the approximate region over which the earthquake was felt. (Intensity values, traditionally indicated with Roman numerals, are shown here in Arabic numerals for simplicity.)

30–40 kilometers across. A typical sand blow in this region was 2–5 meters across and perhaps 10–15 centimeters high, with some as much as an order of magnitude bigger. The area over which liquefaction features occurred at New Madrid is still unprecedented and provides compelling evidence for the enormous magnitudes of the principle events of the New Madrid sequence.

Such data provide evidence that the three largest events in the sequence were the largest, or certainly among the largest, ever witnessed by European settlers in eastern or central North America. The magnitudes of these great quakes may never be established with certainty, and, as I will discuss shortly, they have been the subject of debate in recent decades. Before we delve into this matter, let's consider another question: How often do similar events recur? If the answer is tens of thousands of years, then we might not have to worry too much about similar events happening again any time soon. Or at least we would worry a lot less than if evidence suggested that they happen every few hundred years.

As recently as the late 1980s, Earth scientists grappled with this issue without much direct evidence. Were the great New Madrid earthquakes characterized by a recurrence interval of 50,000 years? Or 5,000 years? Or 500 years? If we consider the current rate of small earthquakes in the New Madrid zone and extrapolate using the typical magnitude distribution, we obtain a recurrence interval on the order of a few thousand years. This approach is fraught with uncertainty, however, because we know that it would fail miserably for some fault segments, including the major segments of the San Andreas Fault. That is, the main locked sections of this fault generate virtually no small earthquakes, so any extrapolation based on the small event rate would vastly underpredict the rate of the largest expected quakes.

To obtain an answer in which we can place more confidence, we must use a tool in the geologist's bag of tricks called paleoseismology or, more specifically in the case of the New Madrid zone, paleoliquefaction. Paleoseismology involves digging into the earth's shallow layers to reveal evidence of previous large earthquakes in the geologic strata. Paleoseismology can involve several different types of field studies. The methods that have been brought to bear on active faults (to be discussed shortly) cannot be used at New Madrid: the faults in this zone are too deeply buried. But recall the massive scale of liquefaction during the 1811–1812 events. This process resulted in surficial geologic features that have permanently altered the landscape within the Mississippi embayment. As a geologist once observed, you can scarcely put a shovel into the ground without hitting a sand blow formed during the 1811–1812 sequence.

Might there perhaps be a layer, or layers, of equally impressive features underneath, the result of earlier earthquakes? This question has proven surprisingly difficult to answer. In general, looking systematically at the ground even a few meters below the surface is difficult. The occasional road cut or a river channel offers only a limited peek into the subsurface crust. The view can be even more obscured by modern sedimentation, vegetation, and erosion. To get a clear look into the crust requires specially dug holes or long trenches, neither of which is easy or cheap to dig.

Early searches for paleoliquefaction features within the Mississippi embayment came up dry, and researchers concluded that events like those in 1811–1812 must be characterized by recurrence intervals at least tens of thousands of years long. To a geologist, this is an unsatisfying answer. If an event is that rare throughout history, it is highly unlikely to have happened during our limited historical record. Such an eventuality is clearly not impossible, though, and the Earth science community accepted the evidence, or lack thereof, unearthed by the early studies.

In the mid-1990s, geologists Martitia Tuttle and Buddy Schweig conclusively identified the smoking gun for the New Madrid region: subtle but clear liquefaction features that were shown to predate the 1811–1812 events. This discovery opened the doors to a modest geologic gold rush. Within a decade, geologists had scraped and found several dozen separate paleoliquefaction sites that evinced several large prehistoric earthquakes in the last few thousand years. Because certain paleoliquefaction features correlate over a broad area, at least two of the prehistoric events—one around A.D. 900 and a second around A.D. 1400–1500—are inferred to have been as large as the 1811–1812 earthquakes.

Interpreting these data is less straightforward than one might wish. Questions and complications abound in relation to the inescapable imprecision of carbon-14 dates, and inherent limitations are always involved in the interpretation of paleoseismic results. For example, even though these is clear evidence for large earthquakes at the southern and the northern ends of the New Madrid zone and their estimated carbon-14 dates overlap, how do we distinguish between a single huge earthquake and two smaller earthquakes that happened 50 years apart?

Such vexing questions plague virtually all paleoseismic data interpretation, but some clues can help to render answers. At New Madrid, the nature of the individual liquefaction features can supply some information about the size of the causative earthquake. Bigger earthquakes generate bigger liquefaction features, and at least some of the prehistoric features at New Madrid are comparable in scale to the ones generated by the 1811–1812 events. There is even evidence suggesting that these earlier events were also extended sequences of two to three large events, just like those of 1811–1812.

The current views of earthquake recurrence intervals in the New Madrid zone may continue to evolve as the painstaking paleoseismic investigations continue. The best available evidence suggests a recurrence intervals on the order of 500–1,000 years for events like those in 1811–1812. With an average recurrence interval in this range, our odds of experiencing a great earthquake given a 200-year historical record are somewhere between 20–40 percent. With odds this high, the fact that we have witnessed the 1811–1812 sequence no longer assaults geologists' sensibilities.

We know, therefore, that the New Madrid zone can generate large earthquakes, and we know about how often they have occurred over the last few thousand years. Now we can tackle the question of just how large these events really were. Modern analyses have obtained estimates ranging from the low M7s to above M8. Were the three largest 1811–1812 events all around M7–7.5? Or were they M8+ earthquakes? In either case, they were clearly enormous events that would be horrendously destructive were they to recur now. But the difference between a M7.5 event and a M8.1 is enormous, both in its own right and in terms of its implications for long-term hazard assessment.

The answer to the question of how big the 1811–1812 New Madrid events were has proven complex beyond all expectation. It has engendered a debate among geologists and seismologists that has been remarkable for the volume of compelling evidence on both sides.

First, we have the substantial collection of reported effects from the 1811–1812 events that, according to one interpretation, suggests magnitudes upwards of M8 for the three largest shocks. The evidence includes the vast extent of liquefaction features in the immediate New Madrid zone—the area over which the earthquake was felt strongly—and observations (such as the sloshing of water within lakes) that suggest the kind of long-period energy that would be generated only by huge earthquakes.

The most-direct evidence that constrains magnitude is what we have gleaned from original sources in which the earthquake effects are described, sources used to determine Modified Mercalli Intensities. For the New Madrid events, Otto Nuttli published a landmark study in 1973 in which he evaluated intensities from a collection of approximately fifty original sources. A decade later, Ronald Street did another extensive archival search and nearly doubled the number of available reports.

Much of what we now know about the size and details of the New Madrid events can be attributed to the exhaustive analysis of seismologist Arch Johnston. Johnston compared the effects of the 1811–1812 events with those of earthquakes in similar geologic terrain worldwide and concluded that the former stand in a class by themselves. Since instrumental magnitudes exist for comparable events elsewhere in the world, one can obtain a calibration with which intensity values can be used to determine magnitude. On the basis of an evaluation of all pertinent data, Johnston obtained magnitude values of 7.8–8.1 for the three largest events.

Johnston's contribution does not stop there, however. By meticulously evaluating the detailed felt reports from New Madrid and elsewhere, Johnston constructed an actual fault rupture scenario for the four largest New Madrid events. Looking at the three main limbs of the New Madrid fault zone (see Figure 7.5), Johnston concluded that the first event ruptured the southernmost strike-slip segment, the January 23 event ruptured the northernmost segment (also strike-slip), and the February 7 event ruptured a thrust fault segment that links the two. The large aftershock at dawn on December 16 (considered the fourth largest event of the sequence) is inferred to have ruptured the northern end of the southern segment. This inference is based on reports from the small town of Little Prairie, located at the northernmost end of the southern segment, that the aftershock intensity at that location was at least equal to that of the first mainshock.

This analysis represents detective work worthy of Hercule Poirot. The fault interpretations are guided by modern seismicity, which is assumed to illuminate the main fault segments of the New Madrid seismic zone. But because the faults are buried under the thick sediments that overlie the rift zone, there is neither surficial data to constrain the fault characteristics now nor any direct observation of surface faulting features from the 1811–1812 earthquakes. Only a couple of intriguing observations suggest any disruption of the earth's surface: reports from a handful of boatmen who noted waterfalls forming on the Mississippi River following the February event. The creation of waterfalls by thrust earthquakes is rare, but it does happen. In 1999, the M7.6 Taiwan earthquake created a spectacular and well-documented waterfall in a river (Figure 7.6).

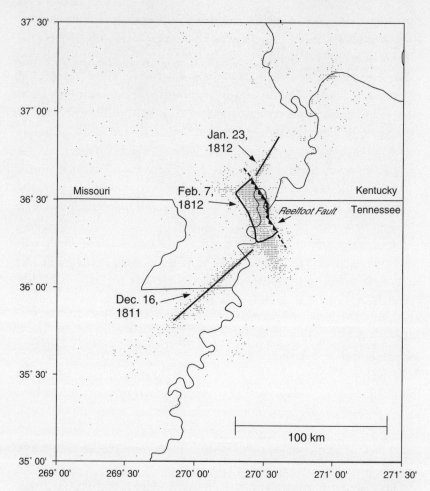

Figure 7.5. The New Madrid seismic zone, as illuminated by small earthquakes (small circles) recorded in recent years. The three faults thought to have ruptured in 1811–1812 are also shown.

The great New Madrid earthquakes altered the landscape in a significant manner, resulting in uplift of some regions and subsidence of others. But imagine pushing a giant heap of sand with a few erratic passes of a bulldozer and trying to figure out the bulldozer's path by looking at the mess it left behind. Now imagine that the sand heap covers the earth's surface and the bulldozer is pushing from perhaps 5 kilometers underground. When the earth's deep crust

Figure 7.6. Waterfall formed in the Tachia River during the 1999 M7.6 thrust earthquake in Taiwan. Following the New Madrid event of February 7, 1812, boatmen reported witnessing the creation of waterfalls on the Mississippi River. Because the New Madrid rupture occurred in a region with heavy near-surface sedimentation, a waterfall such as this one would have been short-lived owing to rapid erosion of the underwater fault scarp. Photograph by Katsuyuki Abe. Used with permission.

was torn asunder at New Madrid, the sedimentary landscape of the Mississippi embayment was left in a shambles.

Johnston's reconstruction of a detailed faulting scenario for the New Madrid sequence is a remarkable achievement indeed, and one is not usually impelled to argue with results that are as amazing as they are scientifically thorough. However, some tenets of seismology appear hard to reconcile with the magnitudes that Johnston infers. To produce big earthquakes, you need big faults. The depth extent of the three major New Madrid earthquakes is unknown, but it is almost certainly less than 30 kilometers. Estimates of the lengths of the faults that generated the events range from 55 kilometers to, at most, 140 kilometers. These estimates yield an available fault area of 1,650–4,200 square kilometers for each rupture. According to well-established relations between fault area and magnitude, these estimates correspond to a magnitude range of approximately 7.3–7.7. Even then, many seismologists argue that the depth of faults in the New Madrid Zone is closer to 15 kilometers than 30 kilometers, which would further reduce the maximum possible magnitudes.

There is one wrinkle in this analysis: a given fault rupture area does not always produce the same magnitude. Recall that some earthquakes have high magnitudes for a given area; seismologists know these as high-stress-drop events. There is, moreover, some evidence, though not conclusive, that earthquakes in the older, stronger crust of the eastern United States have higher stress drops on average than their California counterparts. But even if we assume a stress drop twice as high as the range for large California earthquakes, we still come up with a magnitude range of M7.5–7.9, given the fault geometry at New Madrid.

To account for three M7.8–8.1 earthquakes with the geometry of the three largest New Madrid earthquakes requires a stress drop approximately four to eight times the value consistently observed for the largest events in California and for moderate to large earthquakes elsewhere in the world. Magnitude 8+ earthquakes with enormous stress-drop values defy expectations and sensibilities. Although the stress drop of an eastern earthquake might be higher on average than that of a California earthquake, analysis of the handful of M5–6 earthquakes recorded by modern instruments in the eastern United States and Canada does not support a fivefold difference.

Moreover, the careful analysis by Johnston and others of the maximum earthquake magnitude observed in all parts of the world where the crust can be considered similar (that is, other failed rifts in intraplate crust) reveals that M8.1 is the largest value ever observed for these zones. In any collection of objects of differing size, one object must be the largest. In any collection of earthquakes, one will have the highest stress drop. But to conclude that the New Madrid events were so exceptional is as unsatisfying to an Earth scientist as the conclusion that they happen only once in a proverbial blue moon (but just happen to have occurred since Europeans settled the continent).

The debate therefore continues (Sidebar 7.2). To prove that the 55- to 140-kilometer fault segments could not possibly produce M8.1 earthquakes would be to prove a negative, which is generally not a fruitful endeavor. The debate over the magnitudes of the 1811–1812 events should not, however, obscure the most salient conclusion, which is that all three events were enormous. If the February hard shock was only M7.4, it was still the same size as the 1999 Turkey earthquake, which claimed nearly twenty thousand lives.

One indirect bit of support for the possibility of huge earthquakes in the intraplate crust came from an unlikely source in 1998, when a well-recorded

Sidebar 7.2 New Madrid: The Latest Chapter

Scientists tend to view the writing of books for a general audience rather than for Earth science specialists as an activity that takes time away from our "real work," the research that pushes the edge of the envelope. But in one delightfully serendipitous instance, research for this chapter led me into a research project that I would not otherwise have undertaken. Going back to original accounts of the 1811–1812 New Madrid earthquakes, I stumbled across instances in which the original interpretation of the account had not fully accounted for the phenomenon of site response—the amplification of earthquake waves within soft sediments. Where did the people of the central United States live in 1811–1812? They lived, overwhelmingly, along major waterways. Yet coastlines and riverbanks are notorious for high site response, and some observant eyewitnesses explicitly commented that shaking had been higher along rivers—or in valleys—than elsewhere. Reinterpreting the original reports with a consideration of site response and other issues, my colleagues and I obtained magnitude estimates in the 7.0–7.5 range, estimates more consistent with other lines of evidence. This new study is almost certainly not the last word but rather the beginning of a dialogue that may reshape the debate. And it is a dialogue that would not have begun—at least not at this time and in this way—if not for an effort to communicate the lessons, and passions, of the Earth sciences to a general audience. It is said that to learn, we must teach. I am inclined to amend this old saying a bit: we must teach at all levels. If the gist of an explanation—no matter how complex—cannot be successfully communicated in nontechnical terms, then perhaps the explanation isn't as right as we think.

M8.2 earthquake occurred in oceanic intraplate crust near Antarctica. Like intraplate continental crust, oceanic intraplate crust is stable and earthquake-free compared to the active plate-boundary regions. The oceanic crust is also thinner than continental crust, and so it is generally less able to generate huge earthquakes. But if intraplate oceanic crust can produce a M8.2 earthquake, then why not intraplate continental crust?

The limitations imposed by our historical and instrumental records frustrate Earth scientists and confound our efforts to quantify future long-term hazard. But by developing paleoseismology tools and ingenious methods with which to interpret historical earthquake reports, Earth scientists have opened up a whole new window onto the enigmatic processes that played out in the New Madrid seismic zone.

FINDING FAULT IN CALIFORNIA

Earth scientists in California have it easy in some ways. Want to know what earthquakes are like? Put seismometers in the ground, and you'll record a bunch. The rate of large events might wax and wane at times, but in the twentieth century a magnitude 6 or greater earthquake has struck somewhere in California once every few years.

Californians can be hopelessly jaded sometimes when it comes to earthquakes. The Big Bear aftershock that hit 3 hours after the M7.3 1992 Landers earthquake was considered to be just a rather large aftershock, yet it was a M6.5 earthquake in its own right. One might venture to guess that a M6.5 earthquake elsewhere in the country would be big news.

It is true that Alaska has more and bigger earthquakes than California does, but several factors have conspired to confer on California the designation of "national earthquake field laboratory." The most obvious factor, beyond seismicity rates, is population density: California might not face the highest earthquake hazard, but it faces the highest earthquake risk within the United States. The state also enjoys the simple advantages being physically connected to the rest of the country—which makes it easier for scientists from other areas to visit and work there—not to mention weather that is often delightfully conducive to fieldwork.

The final factor that contributes to the allure of the Golden State, as far as earthquake research is concerned, is perhaps the most critical: its plate-boundary system is on dry land instead of under water. The San Andreas Fault system is one of the most significant plate boundaries anywhere on the planet that can claim this distinction. Better still, much of the San Andreas Fault and the lesser faults of the plate-boundary system run through sparsely populated desert regions. Not only do we not have to do geology underwater, we don't even have to do geology under vegetation. In urban parts of California, we sometimes have to try to do geology under pavement, but even that is less daunting than studying faults offshore.

These factors have facilitated the development of a different type of paleo-seismic research in California. Rather than investigating the nature of pre-historic ground failures, such as liquefaction, we can excavate and investigate the fault traces themselves. This exceptionally young subfield of geology was launched in the mid-1980s by a handful of geologists including then graduate student Kerry Sieh. Sieh spent years conducting painstakingly careful investigations of a trench that was dug across the San Andreas Fault at a place known as Pallett Creek. The creek, located about 60 kilometers northeast of Los Angeles, near the small town of Wrightwood, is yet another place whose name is secure for all time in the Earth science lexicon but otherwise completely obscure. It was here that direct geologic evidence of prehistoric earthquakes on the San Andreas Fault was first uncovered. This line of research would eventually extend the short historical earthquake record in California by many hundreds—even thousands—of years.

Even in this congenial arid setting, finding sites for paleoseismic investigation along the San Andreas is no easy matter. An active strike-slip fault such as the San Andreas lurches every few hundred years during great earthquakes. The motion in each earthquake is lateral; the two sides of the fault slip past each other with virtually no vertical motion. By standing back from the fault and looking at features such as river channels, one can infer the long-term rate of motion, but this represents only the sum of the individual steps contributed by each earthquake. Digging a trench across the fault in an arbitrary spot is very likely to tell you nothing, because there will be no distinct evidence of individual earthquakes.

To conduct paleoseismic investigations of the San Andreas Fault, you need to identify sites where sediments have collected on top of the fault trace at a steady rate for the last 1,000 or 2,000 years. Imagine building a seven-layer cake on top of a strike-slip fault, adding one layer every few hundred years. If an earthquake ruptures the fault after the first layer is done, that layer will be sliced by the earthquake. Now you add the next layer, and perhaps another. Then a second earthquake comes along and rips through the first three layers. And so on. By the time you get to the seventh layer and the fourth or fifth earthquake, you will have tears that each ruptured through several layers and thereby reflect the timing of each earthquake (Figure 7.7).

This is how paleoseismology on the San Andreas is done. The first critical requirement for understanding the chronology of past events is that the layers must have been built in a fairly steady, uninterrupted manner. The second

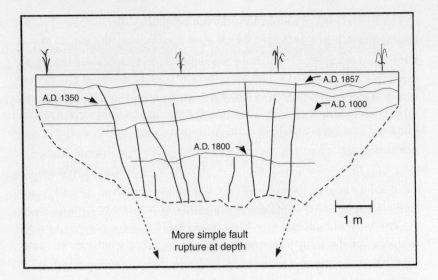

A.D. 1857

A.D. 1350

A.D. 1000

A.D. 1800

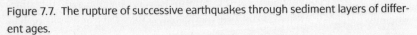

1 m

More simple fault
rupture at depth

Figure 7.7. The rupture of successive earthquakes through sediment layers of different ages.

requirement is that you must be able to deduce the age of each layer. For dating earth materials a few hundred to thousands of years old, the only tool available is carbon-14 dating, which, of course, requires carbon atoms. Within sedimentary layers, carbon is present only where organic materials were trapped, and sometimes it is not easy to find.

At Pallet Creek, Kerry Sieh and his colleagues eventually found evidence of ten separate earthquakes on the fault, eight of them prehistoric. These events appeared to be bunched into four clusters of two to three events apiece. Although carbon-14 dates are never precise, the results suggested that events have occurred in clusters 50–200 years long, separated by quiet intervals of 200–300 years.

Inferring a prehistoric record of earthquakes at Pallet Creek represented a landmark achievement in earthquake science. Yet the years of meticulous analysis required to build this record resulted in but a single snapshot of the 1,000-kilometer-long fault. The observations did constrain the magnitude of the prehistoric ruptures, but we cannot reconstruct a full history of the fault without similar data from other sites along the fault.

By the mid-1990s, a complete view of prehistoric San Andreas Fault earthquakes began to take shape (Figure 7.8). Newer results from additional sites largely confirmed the initial ones from Pallet Creek. Although we can speak of an average recurrence rate of a few hundred years for great earthquakes on the major San Andreas segments, these events do not pop off like clockwork. Instead, the results from other sites confirm the preliminary results that long periods (200–400 years) of quiescence seem to be followed by a few hundred years during which great earthquakes recur perhaps every 100 years on average.

The paleoseismic results from multiple sites along the San Andreas Fault also revealed that earthquakes do not always rupture the exact same segments over and over (as implied by the characteristic earthquake model presented in Chapter 6). The great earthquakes appear to re-rupture the largest segments of the San Andreas, but separate subsegments sometimes fail individually to produce smaller (but still upwards of M7) earthquakes.

With all of the uncertainties that plague geologic and seismologic investigations, there is no substitute for ground truth when it comes to getting Earth science results that inspire a sense of confidence. For the San Andreas, the historical record offers an element of ground truth. That there are sometimes long intervals between earthquakes is beyond dispute; we know with virtual certainty that there have been no great earthquakes on the southernmost segment of the San Andreas for the 300-year interval between 1700 and 2001. We also have direct evidence of two great earthquakes on the Fort Tejon segment within 50 years of each other, the 1857 event and one in 1812 that is documented by written records.

Finally, there is some documentation of the length of the 1857 and 1906 ruptures and the lateral slip at points along the fault. When this data is combined with our knowledge of the depth extent of the San Andreas, we have a modicum of confidence in the magnitude estimation. It is possible to debate such details as whether the average slip of the fault at depth might have been larger than the slip measured at the surface (a not uncommon situation). But the preferred estimate of M7.8 for the Fort Tejon earthquake is uncertain to no more than a few tenths of a unit on either side. Thus are California seismologists spared the kind of detective work with historical records to which eastern seismologists and geologists must resort. (I leave it for the reader to decide which group of scientists gets to have the most fun.)

The body of knowledge amassed for the San Andreas can be translated directly into long-term hazard assessment. The issue is scarcely settled beyond any doubt. Continued work on several fronts—paleoseismic, seismologic, geologic—will continue to provide new insight and possibly improvement in our hazard estimates for some time to come.

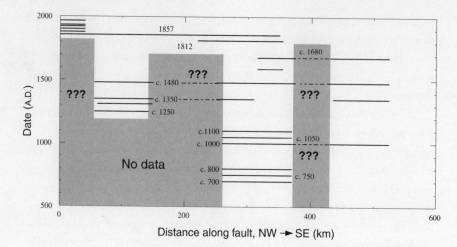

Figure 7.8. Chronology of past earthquakes along the San Andreas Fault, as inferred from paleoseismic investigations. Along large segments of the fault *(shaded regions)* no data are available. Adapted from work by Kerry Sieh, Lisa Grant, and others.

Away from the San Andreas, we have even more to learn. Paleoseismic data can be and have been collected for a growing number of important California faults, including the complicated collection of thrust and strike-slip faults in the greater Los Angeles region and the distributed set of strike-slip faults splintering the Mojave Desert.

Such investigations stand to give new insight into the nature of prehistoric ruptures of California's more minor faults. One example came in 1998, when geologists Charlie Rubin and his colleagues unearthed direct evidence for two M7.5 earthquakes on the Sierra Madre Fault, a major thrust fault just north of Los Angeles, along which the San Gabriel Mountains are being uplifted. Prior to this discovery, there had been considerable debate within the Earth science community regarding the seismic potential of the Sierra Madre Fault. Would it generate earthquakes no bigger than mid- to high-M6 events such as the 1971 Sylmar and 1994 Northridge earthquakes? Or could it rupture in M7.5 events such as the 1952 event that took place in an analogous setting to the north of the San Andreas near Bakersfield? Or could the fault even rupture in conjunction with other adjacent major fault systems to produce a cataclysmic earthquake with magnitude 8 or even 8.5?

Prior to 1998, this issue—the M_{max} question described in Chapter 6—was as heated as debates within the halls of science ever get. Proving that M8 events could never happen on the Sierra Madre fault is as difficult as proving that the New Madrid earthquakes could not have been M8+ events. But hard evidence for prehistoric M7.5 earthquakes does certainly put to rest the claim that nothing larger than M7 is possible. The debate has not been entirely settled, but it has at least been narrowed.

Paleoseismological work is slow, exacting, often uncomfortable, and sometimes even dangerous. Trenches cut across a fault trace tend to be dug through soft sediment layers, and therefore shoring up trenches to prevent collapse can be a nontrivial expense and concern. Work in trenches is rarely comfortable, either, as they are often 3–4 meters deep but no more than 1–2 meters wide.

Another more prosaic factor often makes paleoseismic investigations difficult: obtaining permission to conduct them. Seismologists who deploy portable seismic instrumentation sometimes knock on doors to ask property owners if they might install a seismometer on their back porch or in a shed; paleoseismologists ask people if they would mind having a 20-foot-long trench dug through their yard. Although some trenches have been dug on private property, a geologist will typically seek out public sites such as a Veterans Administration hospital or a county park. This further limits the number of suitable sites, in urban areas especially.

Yet geologists will persevere with their paleoseismic investigations because they are the only way we have to extend our knowledge of earthquakes back in time. They are the only tools we have for learning about prehistoric earthquakes and, thereby, improving our understanding of what is in store for us in the future.

Scientists are constantly absorbed in the things we do not yet understand. Unanswered questions related to earthquake hazard will also inevitably frustrate the public. Yet it is important sometimes to stand back and consider how much we do know, and how we have come to know it.

Imagine being transported to California or New Madrid or the Pacific Northwest 100 years ago and being charged with figuring out what earthquakes have happened here over the last few thousand years and at what rate they can be expected in the future. Most people would be hard-pressed to even identify the major faults (let alone the minor ones). We wouldn't have the first clue how to reconstruct the past. To have scratched around the earth's surface for a few decades and figured out such things as the rate of earthquakes on the San Andreas over the last 1,000 years or the precise date and time of a great earthquake in the Pacific Northwest that happened 300 years ago—this represents a collective achievement that is nothing short of remarkable.

EIGHT   BRINGING THE SCIENCE HOME

Science knows no country.

–LOUIS PASTEUR

Science may transcend political boundaries, but when the rubber meets the road, the application of earthquake science cannot transcend country, or even state. A general understanding of earthquake sources and wave propagation will take hazard assessment efforts just so far. Then we have to look around us and ask some questions. What faults are here? How do seismic waves attenuate in this neck of the woods? What geologic structure is beneath our feet? We must also ask a more prosaic question: What level of preparedness has already been established in the area? After all of the background has been covered and all of the issues have been addressed, a final step remains: we must bring the science home. For the purposes of this chapter, home primarily means North America, where I assume most readers of this book reside.

SIZING UP THE RISK

As I have mentioned in earlier chapters, we cannot, at present, predict earthquakes, but we can predict long-term earthquake rates. In areas such as California, Japan, and Turkey, where earthquake rates are high and where we have had a chance to watch earthquake cycles play out over historical time, a detailed understanding of earthquake processes leads to meaningful probabilistic forecasts over periods of 30–50 years. Hazard maps can identify areas that are reasonably likely to experience strong ground motions over the span of an adult lifetime. The earth can still surprise us with infrequent events, but at active plate boundaries, earthquakes usually happen where we expect them to.

In areas where rates are lower, hazard mapping is possible, but over decades, mapping is much less capable of pinpointing future activity. In any one low-seismicity region, we can reasonably bet that no damaging earthquake will happen in any one person's lifetime. Having watched earthquakes play out in central and eastern North America and in similar tectonic settings worldwide, some Earth scientists have concluded that intraplate faults experience episodes of seismic activity separated by quiet intervals of 10,000–100,000 years, or more. Under this paradigm, earthquakes will inevitably pop up in places where geologic evidence of faulting is slim to non-existent, places likely to be woefully unprepared for earthquake ground motions.

In seismically active industrialized areas such as California, Japan, and Taiwan, the task of increasing preparedness has been going on for decades. To borrow (and twist) a maxim that has become well known in an entirely different context: earthquakes don't kill people, buildings kill people. Since we have the technology to design structures that withstand virtually any ground motions—or at least all but the most extreme motions—seismic risk can be mitigated to the point where it is significantly reduced. We know what types of buildings kill people: weak ones that have a low resistance to lateral shaking, and inflexible ones.

The skeleton of a well-constructed, wood-frame house turns out to be quite good at withstanding earthquake shaking. A typical frame is braced laterally as well as vertically, and wood is flexible enough to sway without breaking. Traditional wood-frame designs do have a few potential weaknesses: masonry chimneys, cripple wall (that is, raised foundation) construction, and weak coupling between the frame and the foundation (Figure 8.1). But we know how to fix all of these problems. Simple plywood sheeting will shore up vulnerable cripple walls, and metal bracings can be used to bolt a house to its foundation. Chimneys can be reinforced, or constructed with a modular design.

Considerably more problematic are single-family homes constructed entirely of brick or other masonry materials. When subjected to earthquake shaking, the mortar interface between masonry units has little resistance; walls and entire structures can topple like children's blocks. (Ironically, the lessons of the three little pigs have to be reversed where earthquakes are concerned, although for most people straw is generally impractical for other reasons.)

Among large commercial buildings, design issues are rather more complex. The geometric design of a structure must not be weak in relation to lateral motions; this requirement generally demands that buildings be constructed with solid, braced walls around the entire perimeter of their base. Steel-frame buildings are more flexible as a rule than are concrete ones, although the 1994

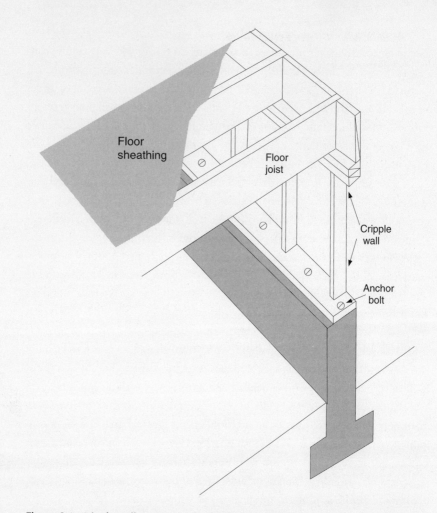

Floor
sheathing

Floor
joist

Cripple
wall

Anchor
bolt

Figure 8.1. Cripple wall construction. Bolting can secure the cripple wall to the foundation and prevent the house from toppling during an earthquake. Covering the cripple walls with plywood paneling can further stabilize the house.

Northridge earthquake highlighted the critical importance of proper welding at beam joints (Sidebar 8.1). Also, concrete structures can be reinforced in such a way as to confer resistance to shaking.

A common design flaw of tall buildings (as far as earthquakes are concerned) is the introduction of a point of differential weakness in the structure, typically one-third to one-half of the way up from the ground. In the absence of earth-

Sidebar 8.1   An Imperfect Union

Steel-frame structures built to meet stringent California building codes are supposed to withstand even strong earthquake shaking. Although steel-frame buildings initially appeared to have sailed through the 1994 Northridge earthquake mostly unscathed, subsequent inspections revealed alarming chinks in the armor. Many of the welds connecting steel beams cracked seriously. Later inspections revealed that ground motions had exceeded the motions anticipated by the building codes, sometimes by more than a factor of two. Many of the failures were attributed to faulty welds, however. Engineers learned that the integrity of highly stressed welds was compromised by even minor imperfections. Within a few short years of the 1994 temblor, state agencies were hard at work taking a new look at building codes and developing new standards for the repair, retrofit, and design of steel-frame buildings. Some day these buildings may well be tested by an earthquake far larger than Northridge. Their occupants will be fortunate indeed that Northridge happened first.

quake hazard, it makes economic sense to build the upper floors of a building less strongly than the lower ones; the latter have to support the weight of the former, whereas the upper floors bear less weight. This practice introduces an unfortunate point of weakness that can become a point of failure when the building is shaken in an earthquake: a single floor can collapse catastrophically in a building that is otherwise left mostly intact (Figure 8.2). Pancaked single stories are a commonplace sight after large earthquakes in urban areas, although, strangely, these buildings often appear undamaged at first glance. Even the modest M6.7 Northridge event caused this type of collapse in one of the few taller office buildings in the immediate epicentral region. Had the earthquake occurred at 4:31 P.M. instead of 4:31 A.M., this one failure could easily have doubled the earthquake's modest death toll. The story was the same when the M7.1 Kobe earthquake shook Japan, a year to the day following Northridge, also very early in the morning.

Like the development of earthquake science, the implementation of earthquake-resistant design is an evolutionary process. In California, building codes

Figure 8.2. A building in which a single floor pancaked during the 1995 M7.1 Kobe, Japan, earthquake.

have been modified and updated through much of the twentieth century. Stringent codes for public schools date back to the 1930s (that is, the aftermath of the 1933 Long Beach earthquake), and freeway retrofitting efforts were stepped up considerably in the 1990s in the aftermath of Northridge. The lessons of the Northridge earthquake were clear: the event caused several catastrophic freeway collapses but no substantive damage to freeways that had already undergone retrofitting.

In Japan, the successful evolution of building codes and practices was illustrated dramatically by the performance of buildings of different vintages during the 1995 Kobe earthquake. Of the buildings constructed prior to World War II, approximately half were rendered unusable in the aftermath of the M7.1 temblor. Yet of structures built since 1971, the corresponding figure was less than 10 percent, and there were almost no instances of catastrophic failure.

In regions where earthquake rates have been low throughout historical times, the problem is clear: seismic design and retrofitting are expensive, and extensive efforts such as those in California and Japan require public resolve. One could scarcely claim the social issues to be trivial, either, at a time when decent, affordable housing is an elusive goal for many.

But a job cannot be completed until it is begun. In 1990, scientists from the U.S. Geological Survey and the Center for Earthquake Research and Information in Memphis summarized a plan of attack for reducing seismic hazard in the Memphis area. Their recommendations included installation of a seismic network, realistic planning scenarios, improved earthquake education and awareness efforts, and initiation of a program to retrofit public buildings. Recognizing the enormity of the last task, the report recommended that efforts begin with the facilities most critical to society: schools, hospitals, and emergency centers.

Although the scientific issues involved in earthquake hazard analysis vary from region to region, it is perhaps true that hazard mitigation knows no country. The blueprint laid out for the Memphis (New Madrid) region could, with minor modification, be applied to virtually any other part of the country not currently well prepared for earthquakes. It bears repeating that we know how to mitigate earthquake risk. We know that we should expect unexpected earthquakes in those parts of North America not considered earthquake country. The question is, how off-guard will we be when the next event occurs?

INTRAPLATE EARTHQUAKES:
LESSONS FROM RECENT HISTORY

We know that unexpected earthquakes should be expected because we have witnessed many of them within the span of our limited historical record. Indeed, we need go back only a decade or two to find examples of surprising events in intraplate regions. On December 28, 1989, a M5.6 earthquake popped up in the city of Newcastle in southeast Australia, not far north of Sydney. This earthquake, the modest size of which would barely merit front-page news in California, claimed thirteen lives and caused a staggering $4 billion in property damage, mostly to older masonry structures. The lessons were clear and sobering to seismologists researching hazard issues in eastern North America. From a geologic perspective, Australia's past is similar to our past; its present could therefore easily be our future.

Oddly enough, the Newcastle earthquake followed almost immediately on the heels of another noteworthy intraplate event, the M6.3 Ungava earthquake, which had struck a sparsely populated part of western Canada on Christmas Day in the same year. Given the temblor's substantial magnitude and shallow depth, its remote location was clearly the only thing that prevented destruction. Heavy snow cover prevented field investigations immediately after the earthquake, but seismologist John Adams and his colleagues at the Canadian Geological Survey assessed the data and concluded that, given the size and depth of the earthquake, it must have ruptured all the way up to the earth's surface—a rare occurrence for an intraplate earthquake. Their supposition was borne out after warmer weather melted the snow and permitted field investigations. Geologists documented a scarp of nearly 2 meters in height on a steeply dipping thrust fault that extended 8.5 kilometers in length across the desolate landscape.

On September 29, 1993, another major intraplate event struck, this one with a devastating combination of appreciable size and unfortunate source location. The M6.1 Killari, India, event occurred not in a remote area but on the densely populated Indian subcontinent. It almost completely destroyed twenty villages in the epicentral region and caused approximately eight thousand deaths, more than three times the number of fatalities as the M7.6 Taiwan earthquake of September 20, 1999. The Taiwan event illustrated the vulnerability of a building inventory that, according to some, had not always been constructed to meet building codes. Still, the death tolls of the Taiwan and

Killari earthquakes did reflect the level of earthquake preparedness in the two countries.

The historical and geologic records tell us that virtually no corner of North America—or similarly quiet places worldwide—is entirely safe from earthquake hazard. Just within the historical record, earthquakes of M5.5—about the size of the Newcastle event—or larger have struck fully half of the states in the Union. In this chapter I examine hazard issues in California, but I also touch on areas in the United States and Canada that are not considered earthquake country. Figure 8.3 serves as a starting point. It shows the expected long-term distribution of shaking in eight major cities within the United States. Although events of all magnitudes up to the regional maximum are expected—and, as

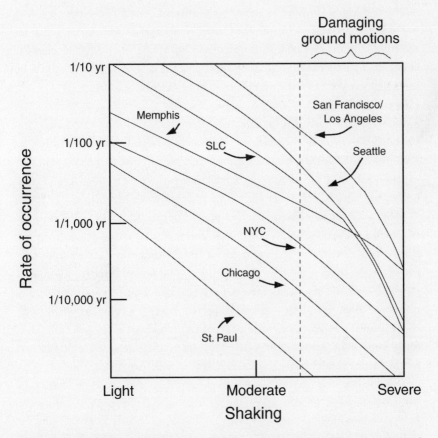

Figure 8.3. The rates of earthquake shaking expected in seven different regions of the United States (SLC = Salt Lake City, Utah).

mentioned, events as small as M5.5 can be damaging—I confine my attention mainly to the largest events that might be expected in a given region, events whose occurrence in the past is known or inferred from direct or indirect evidence. If there is a single overarching principle in geology and seismology alike, it is this: the past is the key to the present, and the future.

HAZARD THROUGHOUT NORTH AMERICA

The Pacific Northwest and British Columbia

There are perhaps parts of the country where earthquake hazard is low, but the Pacific Northwest is not one of them. As far as seismic hazard is concerned, nowhere in the United States has the evolution of understanding been as dramatic, and as unfortunate, as in the far northwest corner of the country. As I noted in the preceding chapter, the Cascadia subduction zone remained so resoundingly mute over historical times as to lull residents and Earth scientists alike into a false sense of security. In retrospect the lack of awareness cannot be too harshly faulted because permanently mute (aseismic) subduction zones do exist elsewhere.

As late as the early 1980s, most geologists and seismologists counted the Cascadia subduction zone among the world's aseismic areas. Within the span of a single decade, however, scientists demonstrated that the region had experienced—and therefore would in the future experience—the largest earthquakes that are likely be seen within the contiguous United States.

What kind of events are we talking about? The proverbial Big Ones, the likes of which have struck South America, Central America, and Japan in recent times, with devastating consequences. If the Cascadia subduction zone ruptures from stem to stern, the fault rupture will be several hundred kilometers long and upwards of 100 kilometers deep. Since earthquake ruptures typically propagate along a fault at a speed of approximately 3 kilometers per second, the fault rupture itself—exclusive of prolonged shaking effects—will continue for as long as 4 minutes. If an event happens to nucleate at the southern end of the zone, residents of Portland, Oregon, will experience shaking that crescendos over the span of a full minute and then dies away gradually over 3 minutes more. In Seattle, shaking may continue to build for a full 3 or 4 minutes before the strongest waves arrive.

Because this will be a subduction event, it will produce significant vertical displacement of real estate, as well as possible tsunami effects. Elastic rebound leads to complex crustal movements following major subduction zone events. Immediate coastal regions are generally pushed upward as the underlying oceanic plate thrusts downward, but significant subsidence can occur inland of the subduction zone. The uplift will create new real estate in areas that were previously underwater. But the subsidence will cause the sea to reclaim some ground as well. Unfortunately, such a rearrangement of the crust is not a zero-sum event for humans, who usually care not about net real estate balance but about their own backyards.

The damage caused by a recent subduction event elsewhere in the Americas—the M8.1 1985 Michoacan, Mexico, earthquake—was devastating, but it stopped short of being truly catastrophic. Although news cameras zeroed in on spectacular instances of building collapse, the dramatic lake bed amplification of waves within the Mexico City lake bed claimed perhaps one out of twenty office buildings. About 95 percent of the buildings in Mexico City remained standing. But in Mexico, the major population centers are not in proximity to the major plate boundary. The major urban centers of the Pacific Northwest, by contrast, are perched immediately atop the subduction zone. And they are poised to pay the price.

If the next great Cascadia subduction event occurs too soon, the price could be steep indeed by virtue of the vagaries of the recent earthquake history. In California the San Andreas fault let loose with two powerful and appropriately menacing growls just as the first European settlers arrived—one in southern California just 8 years after the Gold Rush began and the other in Northern California a half century later. The 1933 Long Beach earthquake provided the heads-up in the Los Angeles region. "You are in earthquake country; build accordingly." But the earth was far less dutiful in warning California's neighbors to the north. Two substantial events happened offshore of Vancouver during the twentieth century, a M7.0 event in 1918 and a M7.3 in 1946. In Washington, the two largest events during the twentieth century were the M6.9 Puget Sound earthquake in 1949 and a M6.5 event, also in the Puget Sound, in 1965. (The 2001 Nisqually earthquake, discussed below, near Seattle was a near twin of the 1949 quake, with a similar magnitude, location, and style of faulting.) The 1949 and 1965 events caused $25 million and $12 million in damage, respectively, and a handful of deaths apiece. However, because the damage was not catastrophic and people were unaware of the enormity of the earthquake hazard posed by the subduction zone, neither these events nor the earlier ones in Canada brought earthquake hazard to the forefront of the public's attention.

As the Pacific Northwest heads into a new millennium, this shining jewel of the Pacific Rim is unprepared for the earthquake hazard that it faces. The M6.8 February 28, 2001, Nisqually earthquake provided a dramatic wake-up call. Although ground motions from the earthquake were relatively low because of the temblor's 50-kilometer depth, it still caused substantial damage to the region's more vulnerable structures. Throughout the city, older unreinforced masonry structures abound, largely defenseless against the lateral shaking of earthquake ground motions. Along the waterfront in Seattle, a double-deck freeway evokes haunting memories of a similar structure that was reduced to rubble in Oakland, California, by the Loma Prieta earthquake of 1989 (see Chapter 4).

As Earth scientists refine our understanding of seismic hazard in the Pacific Northwest—including the threat posed by secondary fault structures associated with the primary subduction zone—the real challenge is a social one. Efforts to remedy the shortcomings in preparedness have begun in Seattle, with projects funded by government and other sources to evaluate vulnerabilities and begin to remedy them. But if you recall that seismic risk can be thought of as hazard divided by preparedness, it is difficult to view Seattle and the surrounding coastal regions with anything but concern.

Northeastern North America

If the paleoseismicity hypothesis is correct—that is, if persistent hot spots of seismicity in the eastern United States are indeed extended aftershock sequences—then an earthquake of approximately M7.2 occurred in central New Hampshire several hundred years ago. An even bigger event, perhaps as large as M7.5, is inferred to have ruptured just offshore of Cape Ann, Massachusetts, a few thousand years ago. Our understanding of active fault structures in the densely populated Northeast leaves much to be desired. Faults on which the rate of earthquakes is low simply do not call attention to themselves in the manner that their West Coast brethren do. Meanwhile, some of the mapped faults do not seem to be associated with the trickle of earthquakes that occurs in northeastern North America. "Earthquakes without faults and faults without earthquakes"—thus has the enigma of East Coast seismicity sometimes been summed up. The experience has been similar in Australia and the stable part of India as well.

If we assume that in the absence of dominant, well-developed fault zones, the long-term rate of big earthquakes can be extrapolated from the rate of smaller events having a standard magnitude distributions, then we can predict the rate of large events expected to occur, somewhere. Using conventional hazard mapping methodologies, we can then calculate the odds of strong ground motions at any place. Such a calculation reveals that damaging ground motions are expected on average every 1,000 years or so in New York City. A calculation for other parts of the northeast, such as Boston and New Hampshire, yields a similar result.

Up into Canada, the level of seismicity in historical times has been a bit higher than the level in the northeastern United States. Several prominent, persistent clusters can be seen in seismicity maps, including along the Saint Lawrence Seaway and around the Charlevoix region of Quebec (Figure 8.4). Within the twentieth century alone, seven earthquakes of M5.7 or larger have struck southeastern Canada. The largest of these, the M7.2 Grand Banks earthquake of 1929, was associated with a submarine landslide, which significantly damaged many kilometers of trans-Atlantic telegraph cables and generated a tsunami. The tsunami—not the earthquake itself—caused a great deal of damage and twenty-nine fatalities in coastal communities of Newfoundland.

Where will the next large East Coast earthquake strike? Perhaps in a remote part of New York State, near the location of the 1983 M5.1 Goodnow earthquake or the 1944 M5.5 Massena earthquake. In spite of the remoteness of these locations, neither possibility is overly benign: no part of the state is deserted. Even the Massena event, which was felt over most of New England, caused $2 million (1944) in property damage. But the next big New York earthquake will be far more devastating if it takes place not upstate but on one of the faults that slice diagonally through Manhattan. One of these faults produced the M4.0 Ardsley, New York, earthquake in 1985; the rupture of others may have generated the sprinkling of M4–5 events inferred to have happened during historical times.

When large earthquakes worldwide cause high death tolls, the American media are often quick to point to the absence of seismic provisions in other countries' building codes and, sometimes, to substandard construction practices elsewhere. The intimation is clear: that won't happen here in the United States. And it is true that if "here" is California, we can conclude fairly that death and destruction on the scale of the 1999 Izmit, Turkey, earthquake are not likely. But what if "here" is Manhattan? Or Boston? Or Washington, D.C., where some of the most famous monuments in the United States sit atop artificially filled land adjacent to the Potomac River?

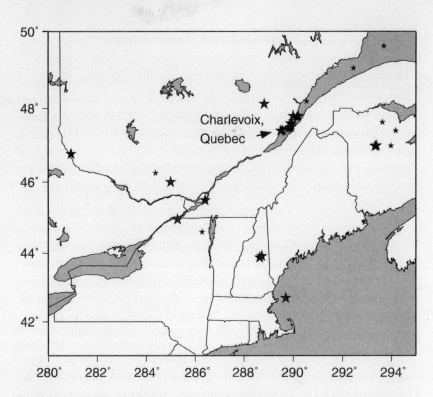

Figure 8.4. Earthquakes in eastern Canada between 1627 and 1998. Large stars are known M5.5+ events; the number of small stars indicates the number of small (M2–3) earthquakes that occur during a given month.

To a Californian, the East Coast building inventory, with its great abundance of unreinforced masonry construction, is almost startling. One good-sized earthquake, one measly 30-second shake, and the richest country in the world could be looking at a disaster the likes of which, one imagines, could never happen in California.

Salt Lake City

Many of Utah's 2 million residents live in the shadow of the Wasatch Mountains. Over three-fifths of the population lives in the region of Salt Lake City and Ogden. Like all mountains, the Wasatch Range exists in a state of balance between tectonic forces pushing them up and inexorable erosion wearing them down. Gently rolling mountains are, as a rule, old mountains, mountains being worn down more rapidly than they are being built up. There is, however, nothing remotely rolling or gentle about the Wasatch Range, whose dramatic peaks provide the setting for some of the best downhill skiing in the country. This mountain range is young; its tectonic processes are very much alive.

The greater Salt Lake area lies within the "Basin and Range" province, where tectonic forces are primarily extensional. In a region with uniformly high elevation, pulling crust apart can result in mountain building as blocks of crust shift and rotate (Figure 8.5). Earthquake hazard in central and western Utah stems primarily from the Wasatch normal fault system, along which the current extensions are concentrated. Unlike in the Pacific Northwest, earthquake hazard in this part of the country was no late-twentieth-century surprise. In 1883—a few short decades after the first Mormon settlers arrived in the Salt Lake area—geologist G. K. Gilbert of the U.S. Geological Survey spoke publicly of the earthquake hazard posed by the Wasatch Fault.

In the hundred years that followed Gilbert's warning, a lot of knowledge was assembled regarding hazard in the Salt Lake area. Geologists identified the half dozen major segments that the 400-kilometer Wasatch Fault comprises. Each segment appears to rupture independently in earthquakes as large as M7.5 (Figure 8.6). Over the past few thousand years, earthquakes of M6.5 and larger have struck approximately once every 350 years on average. To make matters worse, many of the most densely populated urban centers are located on lake bed sediments expected to amplify earthquake shaking.

The Wasatch Fault segment immediately adjacent to Salt Lake City may be one of the most potentially dangerous segments because its last big event was on the order of 1,300 years ago. Most of the other segments have more recently ruptured in large earthquakes. Current hazard models yield appreciable probabilities of a large event on one of the central segments of the Wasatch fault: 13 percent in 50 years and 25 percent in 100 years.

On August 17, 1959, an earthquake that serves as an analog for the next great Wasatch earthquake took place in the southeast corner of Montana, just west of the entrance to Yellowstone National Park. The M7.3 Hebgen Lake earthquake happened not on the Wasatch Fault but rather on another extensional fault system associated with the volcanic processes at Yellowstone (Sidebar 8.2).

Fortunately, the Hebgen Lake region was sparsely populated in 1959, so the death toll from the temblor was a modest one: twenty-eight people were killed, nineteen of whom perished as a consequence of the massive Madison Canyon rock slide. This was no ordinary rock slide; the entire side of a mountain gave

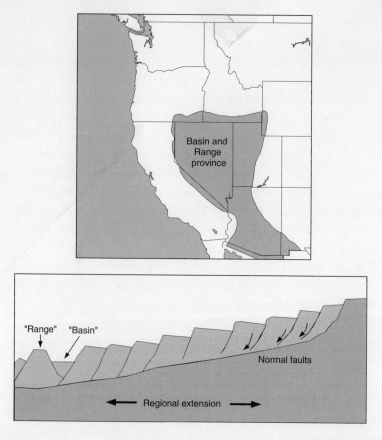

Figure 8.5. Basin and Range. In areas where the crust is uniformly elevated, the process of extension will result in faulting that pulls the crust apart and generates a series of tilted blocks. Mountains (ranges) and basins are the result.

way when the strong earthquake waves fractured and then shattered a nearly vertical layer of strong (dolomite) rock that had buttressed a steep and sizeable mountain slope. When the buttress gave way, some 80 million tons of rock and debris plummeted into Madison Canyon, a scenic valley that was home to the Rock Creek campground. At 11:37 P.M., most of the campers would presumably have been asleep when the earthquake and massive slide struck. What they must have heard and felt that night defies imagination, as does the enormous scale of the disaster. Not only was recovery of bodies out of the question, but

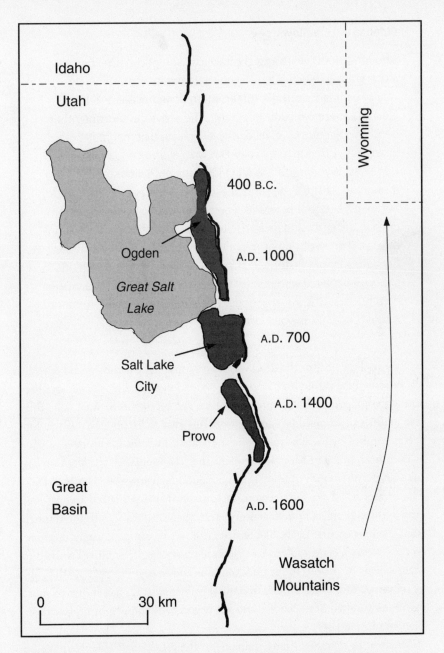

Figure 8.6. Segments of the Wasatch normal fault system in Utah. The dates of the most-recent large events, as inferred from geologic investigations, are shown.

Sidebar 8.2 Yellowstone

About 640,000 years ago, a cataclysmic eruption of Yellowstone volcano ejected 2,000 cubic kilometers of material. By way of comparison, the 1980 eruption of Mount Saint Helens ejected a mere 0.5 cubic kilometer. The active geysers and other geothermal features at Yellowstone indicate that magmatic processes continue to this day. But what is a volcano doing in northwest Wyoming, anyway? Oddly enough, Yellowstone is thought to exist for the same reason that the Hawaiian Islands do: because a hot spot serves as a persistent conduit of magma from deep in the earth and gives rise to vulcanism that lasts for millions of years. For a volcano as impressively alive and bubbly as it has been throughout historical times, Yellowstone has been remarkably well behaved with respect to eruptive activity and even major earthquakes. One has to wonder how long the quiet will continue. Almost certainly, it won't last forever.

even compiling a list of the dead was no easy task. The names would eventually be assembled through a painstaking evaluation of missing-person inquiries and written records such as correspondence and receipts. The final tally: thirteen adults and six children ranging from one year to sixteen years of age, forever entombed by the landscape whose grandeur beckons to this day.

Even now, much of Montana remains sparsely populated, although on any summer day thousands of tourists can be found in Yellowstone National Park. The greater Salt Lake City region, in contrast, has experienced tremendous growth through the twentieth century. And when the next significant Wasatch front temblor occurs, Utah, like Seattle, will not be caught entirely unaware. As the twentieth century drew to a close, Utah state representatives worked to enact seismic provisions in the building code and to begin improving the public infrastructure. In keeping with natural priorities, Utah has evaluated more than four hundred of its schools and has begun to strengthen or replace those deemed substandard.

Much work remains. Many residents of the greater Salt Lake City region live in highly vulnerable masonry homes that predate current building codes. Improvements to such structures are a matter of private, not public, resolve. For a region whose hazard is not too much lower than that of Los Angeles,

San Francisco, or Seattle, Salt Lake City without question still has a long way to go.

New Madrid and the Wabash Valley, Central United States

In earlier chapters I addressed many of the scientific issues related to the quantification of earthquake hazard in the New Madrid seismic zone. We now know, beyond a shadow of a doubt, that this intraplate region can generate enormous and potentially devastating earthquakes with magnitudes upwards of 7. We also know that the thick layer of sediments associated with the Mississippi River embayment will give rise to powerful amplification effects, including liquefaction and slumping along riverbanks.

With respect to hazard, New Madrid rates higher than any other region in the eastern or central United States. So, too, is the level of awareness of earthquake hazard higher there than in many other regions. However, the risk mitigation issues facing the New Madrid region, which includes Memphis, are similar in many ways to the issues in other areas that we have already considered.

The hazard itself is quite real. According to the most recent investigations, the New Madrid seismic zone will generate a sequence of two to three M7–7.5 earthquakes on the order of every 400–500 years. Unless the 1811–1812 activity was grossly anomalous, large events may occur in protracted sequences that play out over months and involve multiple, distinct mainshocks. Perhaps we are fortunate that the last big New Madrid sequence happened as recently as 200 years ago; perhaps we are several hundred years away from the next Big One (or set of Big Ones). But when do we expect the next moderate event? An extrapolation of background seismicity rates predicts that a M6.5 earthquake will occur in the region on average about once every 50 years.

The last event of this size—the M6.7 Charleston, Missouri, earthquake— struck more than 100 years ago, in 1895, and caused damage not only in Missouri but also in parts of Kentucky, Indiana, and Illinois. Although not apocalyptic on the scale of the 1811–1812 sequence, a M6.7 earthquake is comparable in size to the Northridge earthquake, which caused $30 billion of property damage in a region that was supposed to be well-prepared for earthquakes.

In 1968, the largest "New Madrid event" of the twentieth century occurred in south-central Illinois; its magnitude was estimated at 5.3. The quotation marks reflect the fact that although the event was initially assumed to be a New Madrid event, geologists eventually came to realize that its location—in the

Wabash Valley between Illinois and Indiana—is a separate seismic zone, distinct from the New Madrid zone. Evidence from prehistoric liquefaction features has allowed geologists to construct an earthquake history for the Wabash Valley zone, and it includes events of appreciable size. This is not good news for the region, which suffered structural damage from the modest 1968 event as far as 170 kilometers from the source.

Since the Wabash Valley and the New Madrid seismic zones can apparently generate M5.5–6.5 earthquakes at an appreciable clip, the south-central United States should perhaps be most concerned not with the next Really Big One, but rather with the next Pretty Big One.

Charleston, South Carolina

After two widely felt temblors in as many days, an editorial appeared in the *Atlantic Constitution* expressing concern about the potential for future large events along the Atlantic coast, a region that had enjoyed comparative immunity for a long time. The date was August 29 in the year 1886. Two days later, at 9:55 P.M., the Charleston, South Carolina, earthquake produced damaging ground motions over much of the state. The effects of the Charleston earthquake, which had a magnitude now estimated at upwards of 7, were nearly as far-reaching as those of the 1811–1812 New Madrid mainshocks. With remarkable reciprocity, ground motions from the Charleston event were felt in the New Madrid region with almost the same severity as had been ground motions experienced in Charleston during the New Madrid earthquakes.

As I noted in the preface, a singularly unfortunate individual named Paul Pinckney was a boy in Charleston in 1886 and, 20 years later, a young man in San Francisco when the great 1906 earthquake struck. Pinckney recounted his observations and unique perspective in the *San Francisco Chronicle* on May 6, 1906, and one sentence bears repeating: "I do not hesitate to assert that the temblor which wrecked Charleston was more severe than that of April 18 last, and in relative destruction considerably worse."

In fact, all available evidence suggests that the 1906 San Francisco earthquake was the larger of the two events. Recall, however, that damage is not a function of magnitude but rather of ground motions. At Charleston, several factors conspired to bring about especially severe shaking. First, the 1906 San Andreas Fault rupture was several hundred kilometers long, distributing the high-amplitude near-field ground motions over considerable length. A causative fault for the Charleston earthquake has never been identified conclusively; however, the distribution of shaking effects suggests a compact source. Second, set in the coastal plains, Charleston is yet another city where sizeable sediment-induced amplifications are expected. Finally, as mentioned in Chapter 7, some evidence suggests that intraplate earthquakes might be associated with higher stress drops than are their plate-boundary counterparts. For a given magnitude, a higher stress drop translates into more-severe shaking at frequencies damaging to structures.

A description of the net effect is best left to one who experienced it first-hand, so I will return to the words of Paul Pinckney:

> The temblor came lightly with a gentle vibration of the houses as when a cat trots across the floor; but a very few seconds of this and it began to come in sharp jolts and shocks which grew momentarily more violent until buildings were shaking as toys. Frantic with terror, the people rushed from the houses and in doing so many lost their lives from falling chimneys or walls. With one mighty wrench, which did most of the damage, the shock passed. It had been accompanied by a low, rumbling noise, unlike anything ever heard before, and its duration was about one minute.

From a population of only forty-nine thousand, the Charleston earthquake claimed nearly one hundred lives and caused property damage estimated at $5–8 million. As many as seven out of every eight homes were rendered unfit for habitation.

After an extensive archival search in 1987, Leonardo Seeber and John Armbruster compiled more than three thousand intensity reports from 522 separate earthquakes associated with the Charleston sequence: a dozen foreshocks (including the two that inspired the *Atlantic Constitution* editorial) and hundreds of aftershocks with inferred magnitudes of 2.4–5.3. Aftershock epicenters, estimated from patterns of shaking documented by the reports, were strewn over much of the state. The temporal distribution of the sequence was, however, more limited. Seeber and Armbruster showed that, by 1889, the regional seismicity rate had fallen to a modest trickle nearly identical to the modern rate.

As recounted by Paul Pinckney, rebirth was a point of pride for Charleston, a city that, even by 1886, was no stranger to the concept of reconstruction. Within 4 years of the earthquake, the city's population had grown by more than 10 percent. Unfortunately, as the earthquakes retreated and the evidence of strong shaking effects disappeared, so, too, did a large measure of earthquake awareness. A hundred years after the Charleston earthquake, coastal South Carolina—home to nearly a million residents by 1999—had become yet another place in eastern North America where seismic hazard does not loom large in the public consciousness.

Within Earth science circles, however, the memories and lessons of 1886 did not fade away. As is the case for the New Madrid seismic zone, efforts to better understand the Charleston earthquake and seismotectonic setting are inevitably frustrated by the limited and enigmatic data. In many ways we know far less about the seismotectonic setting of the Charleston event than we do about the New Madrid seismic zone, even though we have a greater number of contemporary eyewitness accounts with which to constrain the magnitude of the most recent large event (Figure 8.7).

Yet one avenue of study has proven especially fruitful in the Charleston area: paleoseismology investigations. Indeed, the first documented site of paleoliquefaction in eastern North America was found in Charleston in 1983. By combing coastal plains sediments and analyzing earthquake shaking effects preserved in the geologic strata, geologists have assembled a detailed chronology of prehistoric events. They have shown that Charleston is not one isolated source zone but rather a broad zone that produces sizeable events to both the south and the north of a central segment. All three zones generate earthquakes of comparable magnitude and frequency. Over the past few thousand years, these zones have, together, produced five earthquakes, each roughly comparable to the 1886 event. An average recurrence interval has been estimated at approximately 600 years. When the next one will strike is anyone's guess. But according to state-of-the-art hazard maps, long-term hazard in eastern South Carolina is second in severity only to New Madrid among eastern and central regions. Will the severity of the next great Charleston earthquake exceed that of the next great San Andreas Fault event? Time will tell.

Alaska

No state in the United States has as many earthquakes, or as many large earthquakes, as Alaska, which sits atop one of the Pacific Rim's highly active subduction zones. Like the Cascadia subduction zone, the plate boundary adjacent to Alaska can generate earthquakes that rupture a tremendous distance, both laterally along the plate boundary and vertically down the subducting plate. But

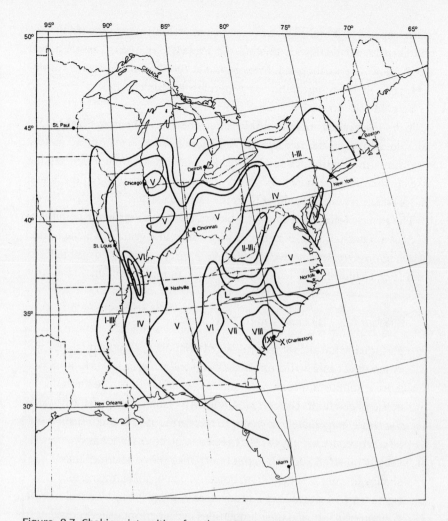

Figure 8.7. Shaking intensities for the 1886 earthquake in Charleston, South Carolina. Adapted from work by Gil Bollinger.[22]

Alaskan subduction not only encompasses a larger lateral extent of plate than does the Cascadia but also is associated with a faster rate of plate motion. As a consequence, earthquake hazard in Alaska has not been a matter of debate. Great tsunami and volcano hazard has long since been recognized as well.

On Good Friday (March 27) of 1964, an earthquake began in the northern Prince William Sound, about 120 kilometers east of Anchorage. It grew into the largest earthquake ever witnessed in modern times in North America; even more impressive, it ranked as the second largest temblor ever recorded by seismic instruments worldwide (behind the 1960 Chilean event, another great subduction zone earthquake). Strong ground motions in Anchorage continued for a full 4–5 minutes and triggered many landslides.

Yet even with shaking of almost unimaginable severity, most of the fatalities (106 of 115) resulted not from shaking effects but from tsunamis (Sidebar 8.3). At Valdez Inlet, a wall of water reached an astonishing height of 67 meters, two-thirds the length of a football field. Fishing boats were tossed around like children's toys (Figure 8.8). Thirty-two people died in Valdez, the largest single concentration of fatalities from the earthquake.

The effects of this tremendous temblor reached much farther than the state of Alaska. Tsunamis of 1–5 meters in height were also generated along coastal Washington, Oregon, and even California. Thirteen people drowned in Crescent City, California, where property damage ran to $10 million.

Sidebar 8.3   By Land and by Sea

Earthquake hazard sometimes arises from unexpected directions. When the oceanic crust experiences a sudden vertical drop—as in a great subduction zone earthquake—water waves are generated at the ocean floor. These waves, known as tsunamis, crash into the shore immediately adjacent to the earthquake source. They also propagate silently across the ocean bottom at speeds of several hundred kilometers per hour. Once they encounter the shallower waters of the ocean bottom at a coastline, their energy becomes concentrated in a progressively more limited vertical column of water, and their amplitudes grow enormously. A wave that cannot be seen in the open ocean suddenly becomes a wall of water, sometimes several tens of meters high. Tsunamis pose a huge hazard to the coastline immediately adjacent to their points of origin, not to mention coastlines many thousands of kilometers away. In the days before warning systems were implemented, ferocious walls of water sometimes crashed onto shores where inhabitants had no knowledge that an earthquake had occurred far away. In Alaska, however, most tsunamis result from local earthquakes and arrive very quickly on the heels of the shaking.

Figure 8.8. Aftermath of the devastating tsunami that struck Valdez, Alaska, following the massive Good Friday earthquake of 1964.

As a direct consequence of the Good Friday earthquake, the West Coast and Alaska Tsunami Warning Center was established in 1967. The center's initial mandate to monitor the coast of Alaska was expanded in 1982 to include British Columbia, as well as Washington, Oregon, and California.

Fortunately, earthquake risk in Alaska is mitigated by its sparse population. Indeed, the 2002 M = 7.9 Denali earthquake caused only limited damage in the wilderness. Unfortunately, investigation of earthquake hazard in Alaska is also limited. Between a harsh environment for fieldwork and the low perceived relevance, support for earthquake science in Alaska has been hard to come by. Yet the population of Alaska—627,000 according to the 2000 census—is one-half to one-third that of states such as Idaho and Utah, states that we scarcely consider deserted. And most of Alaska's population is concentrated along its southern shore, in proximity to the most hazardous earthquake zones. What is more, the population of Alaska swelled by fully 50 percent in the two decades from 1980 to 2000.

By all accounts the dramatic population growth will continue for some time to come; the state offers opportunities that continue to beckon people seeking a new frontier. People who move to Alaska do so with an awareness that the region is a beautiful but inherently challenging place to live. As the newcomers and old-timers shape the Alaska of tomorrow, they must remember that the obvious, day-to-day challenges of life on the northern frontier are only the tip of the iceberg.

Hawaii

Hawaii and Alaska may be vastly disparate settings—geologically and otherwise—but in terms of hazard they are united by a vulnerability to devastating tsunamis. As in Alaska, tsunami hazard in Hawaii represents only part of the geologic hazard. California might be earthquake country, but perhaps no state in the union is more alive in a geologic sense than Hawaii. The very existence of the islands can, after all, be credited to hot spot vulcanism—the continuous upward piping of magma from deep in the earth's mantle. Other forces are at work in Hawaii as well, forces not of creation but of large-scale destruction. As Mauna Loa and Kilauea bring lava to the surface, the Big Island is built up and often steepened. At times the lava heap becomes unstable under the force of gravity, and the volcano's flanks collapse suddenly. In the abrupt lurch of an earthquake, basaltic flanks of the volcano give way and slide toward the sea along large normal faults. The largest such event in recent history, a massive M7.9 temblor in 1923, claimed dozens of lives and many of the structures on the east side of the Big Island. Another event of comparable magnitude happened just a few decades earlier, in 1898.

Like subduction events, flank collapse earthquakes involve substantial vertical displacements and can therefore give rise to tsunamis. The 1898 event caused seventy-seven deaths, forty-six of which resulted from the 12- to 15-meter-high tsunamis that destroyed almost all of the houses and warehouses at Keauhou Landing.

The tsunami hazard in Hawaii and other Pacific Islands is, unfortunately, not restricted to that associated with local earthquakes. Because of the islands' mid-ocean location, the normally idyllic Hawaiian coasts are vulnerable to tsunamis generated by earthquakes at many places along the Pacific Rim, thousands of miles away. The most devastating tsunamis to have struck Hawaii were caused by large subduction earthquakes in the Aleutian Islands in 1946 and off the coast of Chile in 1960. After the Aleutians quake, more than 170 residents of Laupahoehoe and Hilo perished when 10-meter-high waves crashed into the towns with no warning.

In the aftermath of the devastation from the 1946 tsunami, the U.S. Coast and Geodetic Survey established what is now known as the Pacific Tsunami Warning Center, headquartered in Honolulu. By monitoring signals from seismometers worldwide, the center can issue a tsunami watch within 30–40 minutes of an earthquake deemed capable of generating a tsunami. If and when tide stations near the earthquake confirm the generation of tsunamis, the watch becomes a warning accompanied by predicted arrival times at several locations including Hawaii (Figure 8.9). Local authorities can then evacuate low-lying coastal regions, although the height of any tsunami varies drastically in relation to the detailed rupture characteristics and location of the earthquake.

Tsunami warnings are especially effective for distant earthquakes; the tsunamis generated by these earthquakes can take as long as 14 hours to reach Hawaii. For earthquakes originating on Hawaii itself, mitigation and early warning become vastly more difficult. The tsunami generated by a flank collapse will arrive immediately on the heels of the shaking, leaving virtually no time for effective early warning. Mitigation in this circumstance is largely a matter of advising residents of Hawaii's coastal regions to move to higher ground and be alert at the first sign of earthquake shaking.

From a geological perspective, volcanoes are indeed alive: they are born of fire, and they age via processes that involve massive earthquakes and tsunamis. It is small wonder that the Hawaiians of olden times paid homage to volcano gods and goddesses, and took those deities seriously.

Oklahoma

Oklahoma is, oddly enough, home to one of the great modern seismological enigmas of North America, one that is known to few people outside the Earth science community: the Meers Fault. What interests Earth scientists about this major discontinuity of the midcontinental crust is the distinct scarp that is easily visible along one section of the fault. Some 25 kilometers long, the scarp is inferred to have resulted from two big earthquakes within the last few thousand years, the most recent event having taken place perhaps 1,000–1,500 years ago.

Yet the geologic strata reveal no evidence of earthquakes on the Meers Fault prior to the recent ones, the fault has remained stubbornly quiet. In 1985, the Geological Survey of Oklahoma installed a seismograph in a privately owned store. This instrument has proven to be especially sensitive to earthquake waves.

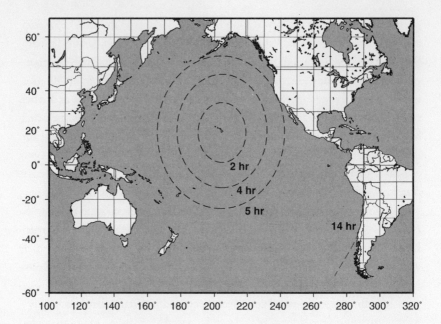

Figure 8.9. Travel times required for tsunamis generated along the Pacific Rim to reach Honolulu, Hawaii.

It has recorded earthquakes from the Indian Ocean and nuclear tests in the former Soviet Union. Yet it was in operation for fully a dozen years before it first recorded an event on the Meers Fault itself: a M2 earthquake in April of 1997.

The Meers Fault may represent the canonical intraplate fault: it has a low rate of activity with active episodes separated by hundreds of thousands of years or more. Apparently faults like this do not always pop off with single events when they do spring to life; instead, they go through episodes of heightened activity that last for thousands or tens of thousands of years. The question for Oklahoma is this: Is the Meers Fault at the end or the middle of its modern episode of activity?

For the rest of the country—and for many other countries—the question is a different one: How many other faults like the Meers Fault are out there, waiting to surprise us? A growing number of Earth scientists suspect that the answer is "a lot." If earthquakes strike infrequently on a particular intraplate fault but we continue to see a steady trickle of intraplate earthquakes in other regions, the inescapable conclusion is that there are scads of faults out there, including a great many we don't yet know about.

Mexico and Central America

The boundary between the North American and Pacific plates changes in character at almost the exact spot that it crosses the border between California and Mexico. Immediately south of the border, there is a zone of active extension that is slowly stripping Baja California from the rest of the continent. Moving farther south, we find an active subduction zone along the coast of Mexico and Central America.

Throughout most of Mexico and Central America, earthquake hazard is akin to that in Alaska and the Pacific Northwest. Coastal regions in this area are therefore exposed to very large earthquakes. In some instances, these subduction zone events pose a hazard for cities and towns away from the coast. The most notorious case, mentioned in an earlier chapter, is Mexico City, where soft sediments in a former lake bed can dramatically amplify ground motions from distant events.

Major subduction zone earthquakes in Mexico and Central America are typically M8–8.5. Those regions face an additional hazard from a different type of earthquake associated with subducting crust: the so-called slab-bending events. Slab-bending earthquakes occur not at the interface between the plates but rather within the subducting oceanic crust. Subjected to bending and gravitational forces, the subducting crust sometimes deforms internally. Although not as potentially huge as subduction zone events, slab-bending earthquakes can still be quite large. One recent, tragic example occurred on January 13, 2001, in El Salvador and caused substantial damage along the country's coast.

Almost all plate boundaries share two critical traits: they generate large earthquakes, and they are far more complicated than scientists have ever predicted on the basis of simple plate tectonics theories.

California

As the country's de facto natural earthquake laboratory, California boasts the most thorough earthquake hazard quantification and highest preparedness of any state in the union. Yet it behooves us to remain mindful of a simple reality: since 1906 California has not been tested by the worst temblors that we know are to be expected on a regular basis.

We have witnessed sobering levels of destruction from several moderate earthquakes that have struck in populated areas since 1906: Long Beach

(1933), Sylmar (1971), Loma Prieta (1989), and Northridge (1994), among others. The price tag for these four events came to about $50 billion even though the state's buildings and freeways performed well (especially for the later events).

The Golden State has gotten to be a crowded place with an enormously expensive building stock and infrastructure. Within the greater Los Angeles area alone, total building inventory amounts to more than half a trillion dollars. The California nightmare is not another Northridge-scale event costing on the order of $30 billion but a Big One that might cost upwards of $100 billion.

Similar concerns exist in two places worldwide with faults like the San Andreas: New Zealand and northern Turkey. In Turkey, the North Anatolian Fault produced several Big Ones in the twentieth century, as noted in Chapter 3. For many decades the Big One in California was assumed to be synonymous with the next great rupture on the San Andreas Fault, such as a repeat of either the 1857 Fort Tejon or 1906 San Francisco event. Toward the end of the twentieth century, however, geologists and seismologists began to recognize the possibility of a different sort of Big One—an event that ruptures not the San Andreas but other large fault systems closer to major urban centers. The closer-to-home event began to loom especially large in Southern California as scientists became more aware of the complex fault systems in the greater Los Angeles region, Orange County, and Santa Barbara.

Although scientists have debated the maximum magnitude possible on the Southern California fault systems (see Chapter 6), the consensus is that a value of M7.5 is appropriate. Although still smaller than the largest San Andreas events, a M7.5 earthquake on a Los Angeles basin thrust fault would certainly qualify as a Big One for the tens of millions of residents of the greater Los Angeles region. It would be approximately twenty times as powerful—with respect to energy release—and five to ten times as long in duration as the Northridge earthquake. The effect might be roughly similar to that of ten Northridge earthquakes laid end to end.

To anyone who witnessed the aftermath of the 1994 temblor in the immediate epicentral region, the prospect of ten sequential Northridge events is truly sobering. But the aftermath could easily be worse still. Although the strongest shaking at any point along a fault will occur as the fault rupture passes by, major shaking will also be generated as the rupture approaches and then moves away from that spot. Thus, while the peak ground motions from a M7.5 event might not be that much higher than from a M6.7, the duration of shaking will be much longer.

In Northern California, meanwhile, the frightening scenario is an earthquake that is twenty times as large as the M6.9 Loma Prieta event and is located much closer to the densely populated San Francisco Bay Area. In Northern California, there is no equivalent to Southern California's San Gabriel Mountain range to buffer the biggest cities from the biggest faults. A repeat of 1906 would run close to the heart of San Francisco itself. In the East Bay, the Hayward Fault is thought to be capable of generating earthquakes upwards of M7 (Figure 8.10), and it runs through the heart of Oakland and Berkeley. Because ground motions immediately adjacent to a fault rupture are known to be especially severe, the hazard associated with the Hayward Fault is substantial because of the high population density immediately along the fault.

In 1999, a working group of experts evaluated the odds that an earthquake of M6.7 or larger—one capable of causing widespread damage—would strike the Bay Area. Their conclusion? There is a 70 percent chance that such an event will strike before 2030. The high hazard stems in part from the San Andreas itself but also from the large collection of secondary faults that splinter off from the San Andreas.

Earthquake hazard in California is by no means restricted to the major population centers of San Francisco and Los Angeles. Hazard maps reveal earthquake threats virtually everywhere, from San Diego (which some scientists have likened, tectonically speaking, to Kobe, Japan) to the far northern reaches of the state. Well north of San Francisco, toward Cape Mendocino, the mighty San Andreas Fault reaches its northern terminus just offshore, in what Earth scientists call a triple junction—a point at which three tectonic plates meet. The Mendocino triple junction marks the boundary between the San Andreas system to the south and the Cascadia subduction zone to the north; the region experiences numerous moderate to large earthquakes.

Is California prepared for the many Big Ones that it ultimately will face? Relatively speaking, the answer is yes. The state's building codes and standards are among the best in the world. Yet the fact remains that California has not been tested by the worst that the earth can produce. California has not faced a Big One since the state emerged as the most populous in the nation, home to a diverse and increasingly high-tech economy.

Sophisticated computer programs offer quantitative estimates of damage, injuries, and fatalities (loss estimation) caused by major earthquakes. Although

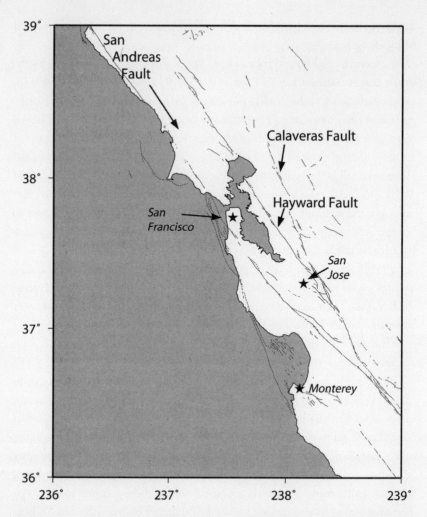

Figure 8.10. Faults in the greater San Francisco Bay Area. The San Andreas Fault runs through the peninsula, posing a high hazard to that region (including San Francisco itself). Other faults splinter off toward the East Bay, menacing the rapidly growing cities of San Jose, Oakland, Berkeley, and Livermore.

very much a developing technology, these programs produce sobering results. For a M7.4 earthquake on the San Andreas Fault north of Los Angeles—an event similar to the 1999 Turkey earthquake—the models predict as many as two thousand fatalities for a daytime event. For a M7.4 temblor on one of the Los Angeles basin faults—a rupture similar to the 1999 Taiwan earthquake—the death toll could be much higher still. In earthquake country complacency is a luxury we can ill afford.

MITIGATING THE RISK

What do we do about earthquake hazard? Not "we the government" but "we the people"? Most of us have little say about the construction of the schools that our children attend or the design of the freeways on which we drive to work. Of course, we the people are we the government, and so societal commitment to mitigate earthquake risk is driven by awareness and concern on the part of the public and the scientific community. But translating scientific and public awareness into public policy is, inevitably, slow going.

Fortunately, many people do have a large measure of control over their own destiny as far as earthquakes are concerned. The earthquake resistance of a single-family home can be evaluated, and many common problems can be alleviated. It is not cheap to reinforce cripple walls, bolt a house to its foundation, or rebuild a chimney, but neither is it exorbitantly expensive. All three reinforcements can be done on a typical single-family home for far less than the cost of installing an in-ground swimming pool. Automatic gas-shutoff valves can be purchased for a few hundred dollars. Putting hard-earned money into improvements that you can't even see, much less enjoy, may seem unsatisfying. Yet what value does one place on peace of mind? These three reinforcements alone can make the difference between a house that looks "blown up" after a major temblor and one that rides out the shaking with minimal damage. A simple gas-shutoff valve can make the difference between a house that goes up in flames when gas lines inside the house rupture and one that does not.

Risk mitigation is also possible within the walls of a home. These efforts are also critical because no matter how well a structure rides out earthquake waves, the interior will be strongly shaken during a large, local event. Fortunately, mitigation of hazards caused by home contents is inexpensive, sometimes as inexpensive as rearranging furniture, art, and mirrors. No matter when and where the next big North American earthquake strikes, the odds are 33 percent that it will happen between 11:00 P.M. and 7:00 A.M. The next time you lie down in bed, take a moment to look around. Anything on or next to the walls immediately adjacent to the bed could end up on the bed—on you—in a heartbeat when the ground begins to shake.

Also, consider the items you would want to have within easy reach if an earthquake knocked out power and damaged your house at 3:00 A.M. Topping the list are a good flashlight (with fresh batteries!) and sturdy shoes.

Emergency supplies will also be critical in the aftermath of a large event: nonperishable food supplies, water, a battery-powered radio, flashlights, first-aid kits, and, where necessary, ample supplies of medication.

Extensive information on personal preparedness is available through many sources, including the Red Cross and, in California, the front pages of the phone book. Structural retrofitting generally requires professional expertise, but such expertise is not hard to find. Beyond the local actions we can take as individuals, global solutions remain a more difficult goal.

Efforts to understand and mitigate earthquake risk—in less well prepared regions especially—will come about when public awareness and concern rise to the point where ill-preparedness no longer is tolerated. A M7.2 earthquake along the coast of Massachusetts or a M6.4 event in New York City would surely do wonders to move things along. But the price of waiting could be steep, and it need not be paid. Advances in Earth sciences have allowed us to quantify seismic hazard and to delineate the shaking expected from future earthquakes. After decades of revolutionary and evolutionary science, we have assembled enough puzzle pieces to give us not the absolute last word on earthquake hazards but a remarkably good working estimate. Parallel advances in building design have taught us how to build resistant structures. A final step remains, however, and it is at least as much a matter of public policy as of science. We need to bring our hard-won lessons home before the earth brings them crashing down on top of us.

NOTES

1. Paul Pinckney, *San Francisco Chronicle,* May 6, 1906. All of the Pinckney quotations in the preface are drawn from this source.

2. Charles Darwin, *The Origin of Species by Means of Natural Selection* (London: J. Murray, 1859).

3. Robert Dean Clark, "J. Tuzo Wilson," Society of Exploration Geophysics Virtual Museum, <http://seg.org/museum/VM/bio_j__tuzo_wilson.html>, October 2001.

4. Dan McKenzie and Bob Parker, "The North Pacific: An Example of Tectonics on a Sphere," *Nature* 216 (1967): 1276–1280.

5. Nancy King, "Back in the Old Days . . . ," *Southern California Earthquake Center Quarterly Newsletter* 5, no. 2 (2000): 3.

6. Arthur Holmes, "Radioactivity and Earth Movements," *Transactions of the Geological Society of Glasgow* 18 (1928): 559–606.

7. James Mori, "Rupture Directivity and Slip Distribution of the M4.3 Foreshock to the 1992 Joshua Tree Earthquake Southern California," *Bulletin of the Seismological Society of America* 86 (1996): 805–810.

8. Douglas Dodge, Gregory Beroza, and William Ellsworth, "Foreshock Sequence of the 1992 Landers, California, Earthquake and Its Implications for Earthquake Nucleation," *Journal of Geophysical Research* 100 (1995): 9865–9880.

9. Robert Wallace, "Eyewitness Account of Surface Faulting during the Earthquake of 28 October, 1983," *Bulletin of the Seismological Society of America* 74 (1984): 1091–1094.

10. Lorenzo Dow, *Lorenzo Dow's Journal* (Wheeling, W. Va.: John B. Wolff, 1849).

11. Yuehua Zeng and John Anderson, "A Composite Source Model of the 1994 Northridge Earthquake Using Genetic Algorithms," *Bulletin of the Seismological Society of America* 86, no. 1b (1996): S71–S83.

12. Timothy Flint, *Recollections of the Last Ten Years, Passed in Occasional Residences and Journeyings in the Valley of the Mississippi* (Boston: Cummings, Hilliard, and Company, 1820).

13. Charles Richter, *Elementary Seismology* (San Francisco: W. H. Freeman, 1958), 66.

14. Ibid., 28.

15. Daniel Drake, *Natural and Statistical View, or Picture of Cincinnati and the Miami Country, Illustrated by Maps, with an Appendix Containing Observations on the Late Earthquakes, the Aurora Borealis, and South-West Wind* (Cincinnati: Looker and Wallace, 1815), 233.

16. David Bowman, G. Ouillon, C. G. Sammis, A. Sornette, and D. Sornette, "An Observational Test of the Critical Earthquake Concept," *Journal of Geophysical Research* 103 (1998): 24359–24372.

17. David Schwartz and Kevin Coppersmith, "Fault Behavior and Characteristic Earthquakes—Examples from the Wasatch and San-Andreas Fault Zones," *Journal of Geophysical Research* 89 (1984): 5681–5698.

18. Steven Wesnousky, "The Gutenberg-Richter or Characteristic Earthquake Distribution, Which Is It?" *Bulletin of the Seismological Society of America* 84 (1994): 1940–1959.

19. Arthur Frankel, C. Mueller, T. Barnhard, D. Perkins, E. V. Leyendecker, N. Dickman, S. Hanson, and M. Hopper, *National Seismic Hazard Maps, June, 1996, Documentation,* U.S. Geological Survey Open-File Report 96-532. Also see "Earthquake Hazards Program—National Seismic Hazard Mapping Project," <http://geohazards.cr.usgs.gov/eq/index.html>, October 2001, for extensive documentation.

20. Working Group on California Earthquake Probabilities, "Seismic Hazards in Southern California: Probable Earthquakes, 1994–2024," *Bulletin of the Seismological Society of America* 86 (1995): 286–291.

21. Jared Brooks's accounts of the New Madrid earthquakes are extensively summarized in Myron Fuller, "The New Madrid Earthquakes," *U.S. Geological Survey Bulletin* 494 (1912).

22. G. A. Bollinger, "Reinterpretation of the Intensity Data for the 1886 Charleston, South Carolina, Earthquake," *Studies Related to the Charleston, South Carolina, Earthquake of 1886—A Preliminary Report,* ed. Douglas W. Rankin, U.S. Geological Survey Professional Paper 1028 (Washington, D.C.: U.S. Government Printing Office, 1977), 17–32.

SUGGESTED READING

ONE THE PLATE TECTONICS REVOLUTION

Kious, W. Jacquelyne, and Robert I. Tilling. *This Dynamic Earth: The Story of Plate Tectonics.* U.S. Dept. of Interior Publication. Washington, D.C.: GPO, 1996. A softcover publication available from the U.S. Government Printing Office. A highly readable and beautifully illustrated introduction to plate tectonics. It is also available on the Web at <http://pubs.usgs.gov/publications/text/dynamic.html>, October 2001.

Keller, Edward A., and Nicholas Pinter. *Active Tectonics: Earthquakes, Uplift, and Landscape.* Upper Saddle River, N.J.: Prentice Hall, 1996. Written at the level of an introductory undergraduate textbook. An in-depth, academic discussion of issues related to earthquakes and plate tectonics.

Moores, Eldridge M., and Robert J. Twiss. *Tectonics.* New York: W. H. Freeman and Co., 1995. An advanced academic treatment of plate tectonics for readers wishing to explore the science in some depth.

TWO SIZING UP EARTHQUAKES

Bolt, Bruce A. *Earthquakes.* 4th ed. New York: W. H. Freeman and Co., 1999. The best available primer on basic seismology. Covers introductory seismology thoroughly without being overly technical.

Richter, Charles F. *Elementary Seismology.* San Francisco: W. H. Freeman and Co., 1958. Out of print but can be found easily in libraries. Predates many of the modern advancements discussed in this book but still remains an invaluable source of basic seismological information.

Keller, Edward A., and Nicholas Pinter. *Active Tectonics: Earthquakes, Uplift, and Landscape.* Upper Saddle River, N.J.: Prentice Hall, 1996.

THREE EARTHQUAKE INTERACTIONS

King, G.C.P., R. S. Stein, and J. Lin. "Static Stress Changes and the Triggering of Earthquakes." *Bulletin of the Seismological Society of America* 84 (1994): 935–953. A fairly technical article in a journal that can be found only in Earth sciences libraries. The first article to carefully lay out the basic tenets of static stress triggering.

Stein, R. S. "The Role of Stress Transfer in Earthquake Occurrence." *Nature* 402 (1999): 6059. A readable article presenting an overview of results on stress triggering.

Hill, D. P., P. A. Reasenberg, A. Michael, W. J. Arabaz, G. Beroza, D. Brumbaugh, J. N. Brune, R. Castro, S. Davis, D. dePolo, W. L. Ellsworth, J. Gomberg, S. Harmsen, L. House, S. M. Jackson, M.J.S. Johnston, L. Jones, R. Keller, S. Malone, L. Munguia, S. Nava, J. C. Pechmann, A. Sanford, R. W. Simpson, R. B. Smith, M. Stark, M. Stickney, A. Vidal, S. Walter, V. Wong, and J. Zollweg. "Seismicity Remotely Triggered by the Magnitude 7.3 Landers, California Earthquake." *Science* 260 (1993): 1617–1623. The paper that first documented the phenomenon of remotely triggered earthquakes. Presents observations of seismicity increases in the aftermath of the 1992 Landers, California, earthquake.

Gomberg, J., P. A. Reasenberg, P. Bodin, and R. A. Harris, "Earthquake Triggering by Seismic Waves following the Landers and Hector Mine Earthquakes." *Nature* 411 (2001): 462–466. Discusses both observational and theoretical results concerning remotely triggered earthquakes, including the (currently leading) theory, which holds that these events are triggered by the shaking from distant large mainshocks.

FOUR GROUND MOTIONS

Bolt, Bruce A. *Earthquakes.* 4th ed. New York: W. H. Freeman and Co., 1999. Contains some discussion of earthquake ground motions and is written at a level accessible to a nonspecialist.

Keller, Edward A., and Nicholas Pinter. *Active Tectonics: Earthquakes, Uplift, and Landscape.* Upper Saddle River, N.J.: Prentice Hall, 1996. Contains some discussion of earthquake ground motions and is written at the level of an undergraduate textbook.

FIVE THE HOLY GRAIL OF EARTHQUAKE PREDICTION

Bolt, Bruce A. *Earthquakes.* 4th ed. New York: W. H. Freeman and Co., 1999.

Bolt, Bruce A. *Earthquakes and Geological Discovery.* Scientific American Library series, no. 46. New York: Scientific American Library, 1993. Covers a range of earthquake-related subjects, including earthquake prediction. More narrative in style than *Earthquakes.*

Yeats, Robert S. *Living with Earthquakes in the Pacific Northwest.* Corvallis, Oreg.: Oregon State University Press, 1998. Focuses on earthquake hazard in the Pacific Northwest and discusses general topics related to earthquake geology, including earthquake prediction.

Reiter, Leon. *Earthquake Hazard Analysis.* New York: Columbia University Press, 1991. Describes basic concepts and the application of seismic hazard assessment, including the scientific issues involved in such calculations.

Yeats, Robert S. *Living with Earthquakes in California: A Survivor's Guide.* Corvallis, Oreg.: Oregon State University Press, 2001. Focuses on issues related to earthquake preparedness in California and discusses the state's earthquake history, the evolution of building codes, and the role of government agencies in emergency response. Gives pointers for making a home earthquake safe.

Yeats, Robert S. *Living with Earthquakes in the Pacific Northwest.* Corvallis, Oreg.: Oregon State University Press, 1998.

Sieh, Kerry E., and Simon Levay. *Earth in Turmoil: Earthquakes, Volcanoes, and Their Impact on Humankind.* New York: W. H. Freeman and Co., 1999. Focuses on the human impact of geologic disasters, exploring the scientific issues associated with hazard mitigation.

SEVEN A JOURNEY BACK IN TIME

McCalpin, James. P., ed. *Paleoseismology.* International geophysics series, vol. 62. San Diego: Academic Press, 1998. One of the few books to provide an overview of the relatively young discipline, covering both techniques and case histories. Written for the professional or serious student.

Yeats, Robert S., Clarence R. Allen, and Kerry E. Sieh. *The Geology of Earthquakes.* New York: Oxford University Press, 1996. Focuses on the geologic causes and effects of earthquakes with an emphasis on inferences gleaned from field observations. More accessible than the McCalpin book.

Sieh, Kerry E., and Simon Levay. *Earth in Turmoil: Earthquakes, Volcanoes, and Their Impact on Humankind.* New York: W. H. Freeman and Co., 1999.

EIGHT BRINGING THE SCIENCE HOME

Collier, Michael. *A Land in Motion: California's San Andreas Fault.* Illustrated by Lawrence Ormsby. Berkeley: University of California Press, 1999. A geologic tour of California's most famous fault, written in an accessible narrative style with dramatic illustrations. Includes some basic information about earthquakes and plate tectonics as well as a detailed discussion of the San Andreas Fault and some of its important historic earthquakes.

Yeats, Robert S. *Living with Earthquakes in California: A Survivor's Guide.* Corvallis, Oreg.: Oregon State University Press, 2001.

Yeats, Robert S. *Living with Earthquakes in the Pacific Northwest.* Corvallis, Oreg.: Oregon State University Press, 1998.

Farley, John E. *Earthquake Fears, Predictions, and Preparations in Mid-America.* Carbondale: Southern Illinois University Press, 1998. Primarily a discussion of the 1990 New Madrid earthquake prediction made by Iben Browning. Explores the public reaction to and long-term impact of the prediction.

INDEX

Page numbers in *italics* indicate illustrations.

California, x, 65, 186; hazard in, 219–23; historic and prehistoric earthquakes in, 186–91; large earthquakes in, 70; preparedness in, 219; southern, 29, 134–35. *See also cities and earthquakes by name*
California Division of Mines and Geology, 157, 158, 159
California Institute of Technology (Caltech), 33
Cambridge University, 10, 12
Canada, 146, 158, 159, 203–4
Canadian Geological Survey, 7, 148, 198
Cape Ann, Massachusetts, 146, 202
Cape Mendocino, California, 221
carbon-14 dating, 179, 188
Carroll, Lewis, 107
cascade model, 40, *41*
Cascadia earthquake of 1700, 171, 173
Cascadia subduction zone. *See* Pacific Northwest
catastrophism, 2–3
Center for Earthquake Research and Information (CERI), 197
chaos theory, 112–13, 120; sand-pile analogue for, 113–14, 120
characteristic earthquake, 138, *139*, 140–41, 189
Charleston, Missouri, earthquake of 1895, 209. *See also* New Madrid, Missouri, earthquakes of 1811–1812
Charleston, South Carolina: hazard in, 210–12; site response in, 105–6
Charleston, South Carolina, earthquake of 1886, ix–xi, 21, 146; aftershocks of, 211; effects of, 210–12, *213;* foreshocks of, 211
Charlevoix, Quebec, 203, *204*
Chicago, Illinois, 199
Chile earthquake of 1960, 83, 168–69, 216
China, 20. *See also* Haicheng, China, earthquake of 1975; Tangshan, China, earthquake of 1976
Coachella Valley. *See* San Andreas Fault
Coalinga, California, earthquake of 1983, 125
coda. *See* earthquake waves
coda-Q. *See* attenuation
computer revolution, 10, 78, 96
Confucius, 131, 164
continental collision, 20. *See also* continental drift; plate tectonics
continental drift, 2–4, 10, 12, 22. *See also* continental collision; plate tectonics
continental rift, 19–20, 173
convection. *See* aesthenosphere; mantle
Coppersmith, Kevin, 138, *139*, 141
core, of Earth, 6, 18
corner frequency, 87

Cox, Allan, 9
cripple wall construction, 193, *194, *223
critical point model, 120–23, *122*
crust, 2; brittleness of, 17, 39, 72; continental, 3–4, 173; deformation of, 15; earthquakes within, 17–18, 39; elasticity of, 27–28, 54, 116; extension of, 205, *206;* faults in, 69; fluids in, 73–75, 109; forces acting on, 13, 35; intraplate, 21, 184–85; magnetism of, 6, 7, *8, 9;* near-surface, *see* site response; oceanic, 5–8, *11,* 13, 18–21, 168–70, 173, 185; plates of, 10, *11,* 17–18, 54; pressure within, 29; rigidity of, 54, 72; subduction of, 54. *See also* plates; plate tectonics

Dalrymple, Brent, 9
damage mechanics, 109
Darwin, Charles, 1, 3, 4
Das, Shamita, 68–69, 72
Day, Stephen, 42
Deng, Jishu, 70, 72, 77
Dieterich, Jim, 74, 77
Dodge, Douglas, 41, 119
Doell, Richard, 9
Dolan, James, 134–35, 143
Drake, Daniel, 93
Duzce earthquake. *See* Turkey (Izmit and Duzce) earthquakes of 1999

earthquakes, 1, 12–14, 17, 20, 26, 32, 51; characteristic, 138, *139,* 140–41, 189; complexity of rupture in, 39–43, 53–55; computer models for, 31–32, 39, 42; cycle of, 120, 132, 138, 166, 192; depth of, 32–33; experience of, 25–26; initiation (nucleation) of, 39–41, 56, 109, 115–16, 119–20; interactions (stress triggering) in, 52–79, 115, 119–20, 161, 163–64; intraplate, 21, 138, *144,* 193, 198–200, 202–4, 218; locating, 15, 43–47, 49; magnitude of, 32–39; magnitude distribution of, 31; prehistoric, 165–91; slip-pulse rupture model of, 41–42; spectrum, *see* spectrum, of seismic signal; stress drop of, 38–39; subevents in, 53. *See also b-*value; earthquake prediction; earthquake waves; seismic hazard assessment; *specific earthquakes by name*
earthquake precursors, 109–11, 120, 123; earthquake wave velocity changes, 110–11; electromagnetic signals and, 109, 111, 126–27; foreshocks as, 111, 118–20; ground water fluctuations and, 111; animal sensitivity to, 110; slip on faults and, 116–17; strain and, 118

earthquake prediction, x, xii, 78–79, 107–30, 131, 161; animal sensitivity and, 110, 119; critical point model for, 120–24, *122;* and earthquake nucleation, 56, 115–20; Haicheng earthquake and, 119–20; intermediate-term forecasting and, 79, 161; Loma Prieta earthquake and, 111; and pseudoscience, 128–30; self-organized criticality, implications of, 112, 114, 120–21, 128; Tangshan earthquake and, 120; tidal forces and, 130; VAN method for, 126–28. *See also* earthquake precursors; Parkfield, California, historic earthquakes

earthquake waves (seismic waves), 33, 43, 74; codas, 89–91; complexity of, 83, *84;* converted basin waves, 102, 103, 105; energy of, 33, 87–88, 94; focusing of, 103–5; frequency of, 48, 85–86, 148; generation of, 43; ground velocity, 87; hazard associated with, 80–81; 147–49; in layered media, 94; modeling of, 96–97, 105–6; P and S waves, 43–45, 49, *84,* 85, 89–90, 102; speed of, 17–18, 43, 110–11; surface waves, 84; teleseismic waves, 14, 46. *See also* ground motions; site response; spectrum, of seismic signal

East African rift zone, 19

Ebel, John, 145–46

Einstein, Albert, 28

elastic rebound theory, 26, *27,* 28, 33, 54, 120, 200–201

electromagnetic signals, 109, 111

Elementary Seismology (Richter), 79

Ellsworth, William, 40–41

El Salvador earthquake of 2001, 219

Elsinore fault, *30*

emergency supplies, 224

empirical Green's function analysis, 86

ENIAC, 10

epicenter, 40

Eurasia, 20–21

Eureka Peak Fault, 172. *See also* Landers, California, earthquake of 1992

failed rift, 20, 145, 174, 184

faults, 13, 26–32, 33, 43–45; and aftershocks, 58; blind thrust, 19; computer models of, 30–32, 42, 116–17; creep on, 28, 29, 32, 168; driving forces and, 13, 35; effects of on landforms, 133, *134;* evaluation of for hazard, 131, 133–47; fluids in, 29; fractal character of, 29–30; gouge zone and, 28, 29, 31, 117; immature (poorly developed), 138, 140, 145; jog in, 62, *63,* 73; locking of, 28–29, 116, 120; mature, 138, 140, *141,* 145; nor-

mal, 19; origin of word, 28; rupture in earthquakes, 17, 33, 36, 39–42; in southern California, 29–30; strike-slip, 19, 25, 134; thrust, 19, 25; transform, 10, *11,* 13, 18–19, 21; types of, 18, *19. See also specific faults by name and region*

Field, Ned, 98, 156

flank collapse, 216–17

Flint, Timothy, 67

fluid diffusion, 68, 73–74

Flushing Meadows, New York, 95, 99, *100*

focal mechanism, 43, *45,* 49

force-balance accelerometer. *See* seismometer, strong motion

foreshock, *40,* 56; distribution of, 56–68, 111–12, 118. *See also* earthquake precursors

Fort Tejon, California, earthquake of 1857, 24, 70, 125, 166–67; magnitude of, 37, 66, 141, 189

Frankel, Art, 91, 101–2, 157

Frasier-Smith, Anthony, 111

friction, 116–17; on faults, 27, 29, 31; incorporation of into computer models, 30–31, 116; between plates, 20–21, 170; and pore pressure, 68; role of in attenuation, 88–89; role of in earthquake triggering, 73, 74, 77; velocity-weakening, 116

Fuller, Myron, 176

Gao, Shangxing, 103

Garlock Fault, 140

gas-shutoff valve, 223

geodesy, 15

Geological Survey of Oklahoma, 217

geology, history of, 1–2; early maps of England, 2

geomagnetism, 7. *See also* magnetism

geometrical spreading, 88

geothermal regions, 75

Geysers, The, 75

Gilbert, G. K., 205

Gilbert, William, 7

Global Positioning System (GPS), 14–17, 110; use for accurate timing, 50

Gobi-Altai, Mongolia, earthquake of 1957, 142

Gomberg, Joan, 74–75

gouge zone, 28, 29, 31, 117

Grand Banks, Canada, earthquake of 1929, 203

Grant, Lisa, 188

Graz, University of, 23

Greece, 126–28

Greenland, 3, 4

ground motions, 38, 54–55, 67, 80–106, 108; accounts of, 171, 172; and damage, 147–51;

ground motions *(continued)*
  extreme values of, 149–51; prediction of,
  147, 159–60; very-long-period, 150. *See also*
  earthquake waves; site response
Gutenberg, Benito, 56
Gutenberg-Richter relation, 56, 59. *See also*
  b-value

Haicheng, China, earthquake of 1975, 119–20
Hanks, Tom, 35, 36, 37
Harris, Ruth, 42, 69, 77
Hauksson, Egill, 64
Hawaii, 8, 208, 216–17; Kilauea and Mauna
  Loa volcanoes in, 216; normal faults in, 216;
  tsunamis and, 216–17
Hawaii earthquake of 1923, 216
Hawking, Steven, 107
Hayward Fault, 221, *222*
Hazard mapping. *See* seismic hazard assessment
Heaton, Tom, 41, 170
Hebgen Lake, Montana, earthquake of 1959,
  205; Madison Canyon rock slide and, 205–6,
  208
Hector Mine, California, earthquake of 1999,
  xii, 65, 71–72, *76;* remotely triggered earth-
  quakes associated with, 75, *76. See also* Lan-
  ders, California, earthquake of 1992
Helmberger, Don, 97, 105
Heng, Cheng, 47
Herrmann, Robert, 91
Hess, Harry Hammond, 4–6, 7, 22, 23
Hill, David, 64
Himalayas, 20
Holmes, Arthur, 4–5, 23
hotspots, 8–9, 18, 208, 216
Hutton, James, 3
hypocenter, 40, 46

Idaho. *See* Borah Peak, Idaho, earthquake of
  1983
Iida, Kumiji, 56
Illinois, 209–10
impedance, 92
India, 20–21. *See also* Bhuj, India, earthquake of
  2001; Killari, India, earthquake of 1993
Indiana, 209–10
induced seismicity, 113–14
InSAR. *See* synthetic aperture radar
Ishimoto, Mishio, 56
Istanbul, Turkey, 71. *See also* Turkey (Izmit and
  Duzce) earthquakes of 1999
Izmit earthquake. *See* Turkey (Izmit and Duzce)
  earthquakes of 1999

Jeffreys, Harold, 10
Johnston, Arch, 145, 181–84
Jones, Lucy, 56, 59, 60
Joshua Tree earthquake of 1992, *40,* 62, *63,* 64,
  119; aftershocks of, 60, 119; foreshock of,
  40, 62

Kagan, Jan, 69
Kanamori, Hiroo, 35, 36, 37, 170
Kanto, Japan, earthquake of 1923, 169
Kentucky, 209
Kern County (Bakersfield), California, earth-
  quake of 1952, 190
Killari, India, earthquake of 1993, 198–99
King, Geoffrey, 69
King, Nancy, 16
Kisslinger, Carl, 60
Knox, Lawana, 42
Kobe, Japan, earthquake of 1995, 92; damage
  in, 195, *196,* 197; precursors of, 111

Lamont-Doherty Geological Observatory (now
  Lamont-Doherty Earth Observatory), 11, 12
Landers, California, earthquake of 1992, 61–65,
  63, 69–76, 186; aftershocks of, 64, 73; eyewitness account of, 172; faults associated
  with, 62, *63,* 121; foreshocks of, 62, 119;
  InSAR image of, 16; remotely triggered
  earthquakes and, 64–65, 75–76; seismogram
  of, *84;* strong motion data and, 102, 105;
  rupture of, *63*
Lesser Caucasus Mountains, 98
Linde, Allan, 75
Lindh, Allan, 117, 124
lithosphere, 18; viscoelastic relaxation of, 72,
  116
lithostatic pressure, 29
Little Prairie, Missouri, 181. *See also* New
  Madrid, Missouri
liquefaction, 147, 176, 179–80, 209
Loma Prieta earthquake of 1989, 24–25, 138,
  161–62, 220–21; damage caused by, 80, *82,*
  83, 99–102, 202; magnitude of, 37; precursors of, 111; regional seismicity prior to, 121,
  122; site response associated with, 92, 99–102
Long Beach earthquake of 1933, xi, 197, 201,
  219–20
Long Valley, California, 61, 75, 130
Los Angeles basin, 103, *104,* 105; building inventory in, 220; compression across, 135;
  faults in, 134, 190, 220; hazard in, 102,
  143–44, 201. *See also* California
loss estimation, 221, 223
Lyell, Charles, 3

magma, 5, 7, 8, 18, 169, 208, 216

magnetic reversals, 6, 9

magnetism: of Earth, 6, 7, 9; of iron, 1–2; of seafloor, 6–7, *8, 9*; study of, 7

magnitude, of earthquakes, 33–34, 37–38, 49; saturation of scales, 34; uncertainty of estimates of, 37–38. *See also* M_{max}; Richter magnitude

mainshock, 38, *40,* 56–58, 61, 69, 71, 74, 77, 111, 146. *See also* earthquakes

Mammoth Lakes, California. *See* Long Valley, California

mantle, 4, 5, 17–18, 23, 72, 173

Mariana subduction zone, 168

Marina district of San Francisco. *See* Loma Prieta earthquake of 1989

Massonnet, Didier, 16

Matthews, Drummond, 7, 8, 9

Maxwell, James, 3

McKenzie, Dan, 10, 11, 12

Meers Fault, 217–18

Memphis, Tennessee, 21, 65, 130, 173; hazard in, 197, 209. *See also* New Madrid, Missouri; New Madrid, Missouri, earthquakes of 1811–1812

Mendel, Gregor, 3

Mexico: hazard in, 219. *See also* Mexico City (Michoacan), Mexico, earthquake of 1985

Mexico City (Michoacan), Mexico, earthquake of 1985, 92, 201; damage caused by, 97, 98, 219

mid-ocean ridges, 5, 7, 10, *13,* 18–21, 169; forces caused by mid-Atlantic Ridge, 174–75

Milne, John, 93

Mississippi embayment, 173, 174, 178. *See also* New Madrid, Missouri; site response

M_{max}, 137, 138–44; in intraplate crust, 145, 184–85; on San Andreas Fault, 220–21; in southern California, 190–91; along subduction zones, 169

modified Mercalli intensity (MMI), 176, *177,* 180

Mohorovicic discontinuity (Moho), 17; subcritical reflections from, 100–101

Mojave Desert, 61, 102, 190; faults in, 121. *See also* Landers, California, earthquake of 1992

moment. *See* seismic moment

moment magnitude, 35–38

moment rate, increasing, 121, *122,* 123

Mongolia, 20

Morgan, Jason, 10

Mori, James, 40

Morley, Lawrence, 7, 8, 9, 22, 23

Nahanni, Canada, earthquake of 1985, 148, 150

National Earthquake Prediction Evaluation Committee (NEPEC), 124

Neptunists, 2

Newcastle, Australia, earthquake of 1989, 198, 199

New England, 146. *See also states by name*

New Hampshire, earthquake of 1638, 146, 202

New Madrid, Missouri, 52, 65; faults in, 178, 183–84; hazard in, *158,* 159, 173–86, 209–10; prehistoric earthquakes and, 146, 178–80; tectonic setting of, 173–75. *See also* New Madrid, Missouri, earthquakes of 1811–1812

New Madrid, Missouri, earthquakes of 1811–1812, 21, 52, 65–67, 175–86, *182;* accounts of, 66–67, 175–76, 180–81; aftershocks of, 60, 175–76; effects of, 52, 66, 176, *177,* 180, 182; liquefaction caused by, 176, 178; magnitudes of, 66, 168, 178–86; Mississippi River, effects on, 67, 181; rupture scenario for, 181, *182,* 183–84; site response in, 93, 99, 106, 185; well-level changes in, 73

New York (city), New York, xi, 199, 203

New York (state), earthquakes in, 203

New Zealand, 220

Nimitz Freeway. *See* Loma Prieta earthquake of 1989

Nisqually, Washington, earthquake of 2001, 88, 201–2

Nomicos, Kostas, 126

normal modes, 83

North America, 9, 21

North Anatolian Fault, 19, 220; triggering of earthquakes on, 71, *72*

northeastern North America: faults in, 202; hazard in, 113, 202–4; tectonic forces affecting, 174–75. *See also* seismic hazard assessment

Northridge, California, earthquake of 1994, xii, 54–55, 144, 156, 190, 220; comparison to Nisqually earthquake, 88, 201–2; damage in Santa Monica in, 103–5, *104;* damage to steel-frame buildings in, 160, 193–94, 195; floor collapse in, 195; freeway damage in, 197; ground motions in Tarzana in, 103; prediction of, 123; site response in, 156; strong motion data, 102–3, 151

nuclear weapons testing, 14

nucleation (initiation), 39–41, 56, 109, 115–16, 119–20

Nur, Amos, 68, 73

Nuttli, Otto, 180

Oakland, California, 80, *82,* 83, 99–100, 221
Oklahoma: Meers Fault, 217–18; Geological
    Survey of, 217
Olsen, Kim, 105, 159
omega-squared model for spectrum of seismic
    signal, 86–88
Omori relation, modified, 59
Ontario Science Centre, 9
On the Origin of Species (Darwin), 3
Orange County, California, 220

Pacific Northwest, 168–73; Cascadia subduc-
    tion zone, 168–71, *169,* 173, 200–201; haz-
    ard in, 200–202; historic earthquakes in,
    168, 201; megathrust and, 170; native Amer-
    ican legends and, 170–71; prehistoric earth-
    quakes in, 171, 173; prehistoric tsunamis in,
    170–71, 172; site response in, 105; tsunamis
    in, 200
Pacific Tsunami Warning Center, 217
paleoliquefaction, 178–80; at Charleston, South
    Carolina, 212. *See also* New Madrid, Mis-
    souri, earthquakes of 1811–1812
paleoseismicity, 145–46
paleoseismology, 133–34, 144–45, 146–47,
    178, 180, 187–91, *188*
Pallet Creek, California, 187–88
Pangea, 3
Parker, Bob, 10, 11, 12
Parkfield, California, historic earthquakes, 117,
    118; prediction experiment and, 124–26,
    130, 161
Pasteur, Louis, 3, 192
peak acceleration. *See* attenuation
Pichon, Xavier, 12
Pinckney, Paul, ix–xi, 210–11
plates, 10, 17–18; boundary between Pacific
    and North American, 137; continental, 10,
    54; driving forces on, 13; oceanic, 10,
    54; rigidity of, 10, 17, 54. *See also* crust; plate
    tectonics
plate tectonics, 1–12, *13,* 17–23, 28, 54, 70,
    77; and earthquakes, 26, 120, 162, 166;
    forces associated with, 173; models for,
    30–32; role of GPS in studying, 16; role of
    seismology in development of, 12–14
Plutonists, 2
Poe, Edgar Allan, 80, 106
Portland, Oregon, 200. *See also* Pacific Northwest
postglacial rebound, 21, 28, 175
precariously balanced rocks, 151, *152*
preparedness, 192, 193
Princeton University, 4, 5, 10, 11
P-wave. *See* earthquake waves

quality factor (Q). *See* attenuation

real-time analysis, 46
Reasenberg, Paul, 56, 60
Red Cross, 224
Reelfoot Rift, 173, *174,* 175; sediments in, 181.
    See also New Madrid, Missouri
Reid, H. F., 26–28, 120
remotely triggered earthquakes, 64–65, 75–77
resonance, 94–95, 98, 154. *See also* site
    response
Richter, Charles, 33–34, 37, 52, 54, 56, 78–79
Richter magnitude, 33–34, 36–37
Ring of Fire, *13,* 21, 168, 170. *See also* plate
    tectonics
risk, 108, 131, 133, 192, 223–24
rocks, precariously balanced, 151, *152*
Rubin, Charlie, 190

Saguenay, Quebec, earthquake of 1988, 60
Saint Lawrence Seaway, 203
Saint Paul, Minnesota, 199
Salt Lake City, Utah, xi; hazard in, 199, 204–9,
    207; setting of, 204–5
Sammis, Charlie, 120
San Andreas Fault, 13, 19, 20, 28, *30,* 62, 71,
    96, *134, 136, 167,* 186–90, 220–21; Big
    Bend and, 135, *136,* 137; Coachella Valley
    and, 142, *167,* 168; Carrizo segment of,
    136, 142; creeping section of, 166; future
    earthquakes on, 108, 151, 161–62, *163,*
    166; and Loma Prieta earthquake, 25; magni-
    tude of earthquakes on, 37, 138–43; at
    Parkfield, 125; prehistoric earthquakes on,
    187–89; slip rate of, 121, 134–35; structure
    of through San Gorgonio Pass, *136,* 142,
    167. *See also* Loma Prieta earthquake of
    1989; Parkfield, California, historic
    earthquakes
sand blow, 176, 178. *See also* liquefaction
San Diego, California, 221
San Fernando earthquake. *See* Sylmar, Califor-
    nia, earthquake of 1971
San Francisco, California, earthquake of 1906, ix,
    x, xi; aftershocks of, 60; magnitude of, 37, 66,
    141, 189, 210; rupture of, 167, 189
San Francisco, California, earthquakes in 1970s
    and 1980s, 121, *122;* faults and, 221, *222;*
    hazard in, 221; Marina district, *see* Loma
    Prieta earthquake of 1989
San Gabriel Mountains, 96, 190
San Gorgonio Pass. *See* San Andreas Fault
San Jacinto Fault, *30*
Santa Barbara, California, 220

Santa Cruz Mountains, 80, 99
SAR. *See* synthetic aperture radar
Satake, Kenji, 171
Savage, Jim, 125
SCEC. *See* Southern California Earthquake Center
scenario earthquakes, 105, 159, 160
Scholz, Chris, 68–69, 72
Schwarz, David, 138, 139, 140
Schweig, Buddy, 179
seafloor topography (bathymetry), 4, 5, 23
seafloor spreading, 5, 7; and transform faults, 10, *11,* 12. *See also* continental drift; plate tectonics
seamount, 20
Seattle, Washington, 199, 200, 202. *See also* Nisqually, Washington, earthquake of 2001; Pacific Northwest
Seattle Fault, 121
sediments, 92; and ambient noise, 99; coastal plains and, 212; effect of on ground motions, 92–96, 153–54, 201; mud, 80, *82,* 83; in New Madrid region, 99, 181; oceanic, 170; in San Francisco Bay area, 99; seismic velocities in, 153–54. *See also* site response
Seeber, Leonardo, 71, 211
seismic gap theory, 162. *See also* earthquake, cycle
seismic hazard assessment, 80–81, 108, 131–64, 192, 223; in California, 96, 134–44, 157, *158,* 159, 180; in central North America, 145–47; deterministic, 157, 159–60; in eastern North America, 140–41, 145–47; exceedance levels and, 132; in major U.S. cities, 198, *199;* mitigation and, 133; probabilistic, 131–33, 143, 157; in stable continental (intraplate) regions, 145–47, 192–93; time-dependent, 161; in United States, 157, *158,* 159
seismic moment, 35
seismogram, 81, 83, *84,* 89, *90,* 168
seismometer, 14–15, 33, 34, 47–51, 81–82, 101; array deployment of, 101–2; broadband, 50–51; digital, 48; design of, 47–51; earliest, 47; strong-motion, 42, 49–50, 147; weak-motion, 49–50
self-organized criticality (SOC). *See* earthquake prediction
Sieh, Kerry, 187, 188
Sierra Madre Fault, 190, 191
Silicon Valley. *See* Loma Prieta earthquake of 1989
Simpson, Robert, 69
Singh, Sudarshan, 91

single-scattering model, 89. *See also* attenuation; coda-Q
site classifications, 153–54
site response, 92–95; ambient noise method of investigation of, 98–99; amplification factors and, 153–55; in Armenia, 98–99; in basins and valleys, 92, *93,* 96–97, 101–2, 105, 153; in California, 155; in central and eastern North America, 95, 105–6, 155; in Charleston, South Carolina, 211; and hazard assessment, 153–57, 159–60; in Mexico City, 98; in Mississippi embayment, 99, 209; modeling of, in two and three dimensions, 96–97, 105, 159; nonlinearity of, 155–56
slip, on faults. *See* faults
slip-pulse model. *See* earthquakes
Smith, William, 2
Snider, Antonio, 2, 22
SOC (self-organized criticality). *See* earthquake prediction
Somerville, Paul, 100–1
South America, 2, 21
South Carolina, 158, 159. *See also* Charleston, South Carolina
southern California, faults in, 29, *30,* 142. *See also* faults; seismic hazard assessment
Southern California Earthquake Center (SCEC), 102
Southern California Seismic Network, 62
spectrum, of seismic signal, *85,* 86–88, 91; and damage, 148, *149;* omega-squared model for, 86–88
Spitak, Armenia, earthquake of 1988, 98; damage in Gyumri (Leninikan), 98–99
stable continental regions, 145, 184–85
Stanford University, 9, 101
Stein, Ross, 71, 72
strain, 27–28, 32, 54; budget associated with earthquakes, 135–36; dynamic, 73; precursory, 117
Street, Ronald, 181
stress, 26–29, 32; dynamic changes in, associated with earthquakes, 74–76, 119; and earthquake initiation, 41; and induced earthquakes, 114; normal, 68–70; shear, 68–69; static, 74–76, 119; stress shadows, 69–70, 77, 163–64; transfer of by earthquakes, 53, 68–73; triggering theories of, 68–79
stress drop, 35, 38–39; of intraplate earthquakes, 184, 211
stress triggering theories, 68–79
strong-motion database, 149, 156
strong motion instrument, 42, 49–50, 147